T0214875

Hyperspherical Harmonics Expansion Techniques

Theoretical and Mathematical Physics

The series founded in 1975 and formerly (until 2005) entitled *Texts and Monographs in Physics* (TMP) publishes high-level monographs in theoretical and mathematical physics. The change of title to *Theoretical and Mathematical Physics* (TMP) signals that the series is a suitable publication platform for both the mathematical and the theoretical physicist. The wider scope of the series is reflected by the composition of the editorial board, comprising both physicists and mathematicians.

The books, written in a didactic style and containing a certain amount of elementary background material, bridge the gap between advanced textbooks and research monographs. They can thus serve as basis for advanced studies, not only for lectures and seminars at graduate level, but also for scientists entering a field of research.

More information about this series at http://www.springer.com/series/720

Tapan Kumar Das

Hyperspherical Harmonics Expansion Techniques

Application to Problems in Physics

 Springer

Tapan Kumar Das
Department of Physics
University of Calcutta
Kolkata, West Bengal
India

ISSN 1864-5879   ISSN 1864-5887   (electronic)
Theoretical and Mathematical Physics
ISBN 978-81-322-3793-8   ISBN 978-81-322-2361-0   (eBook)
DOI 10.1007/978-81-322-2361-0

Springer New Delhi Heidelberg New York Dordrecht London
© Springer India 2016
Softcover re-print of the Hardcover 1st edition 2016

Printed on acid-free paper

Springer (India) Pvt. Ltd. is part of Springer Science+Business Media (www.springer.com)

Preface

A major part of research in physics involves solving the Schrödinger equation. While one-body motion in a potential field and a two-body system with mutual interaction are subject matters of standard texts of Quantum Mechanics, an *ab initio* formal solution of the many-body Schrödinger equation for interacting many-body systems is not commonly encountered. The reason for this is that mathematical complexity increases enormously as the number of particles increases from two to three. It is not just the difficulty arising from the increasing number of position coordinates, but also the difficulty in imposing the desired symmetry of the system, identification of appropriate conserved quantum numbers, etc. Naturally, physicists tend to depend on *approximate many-body techniques* e.g. Born-Oppenheimer approximation, variational and perturbation techniques, mean-field theories like Hartree-Fock and Hartree-Fock-Bogoliubov methods, etc. or on the use of models, e.g., shell, collective or liquid drop models in Nuclear Physics. However, a number of problems involving systems containing *a few* particles demand description in terms of coordinates of individual particles. In such cases it is necessary to handle the few-body Schrödinger equation in an exact manner. Hyperspherical harmonics is the appropriate basis for this. With developments in mathematical and computational tools, it is becoming increasingly easy to handle the hyperspherical harmonics basis. Hence it is becoming popular as an effective tool in theoretical research. The hypespherical technique is quite handy for use in the essentially exact Monte Carlo methods for a fairly large number of interacting particles. Unfortunately, there is a dearth of monographs dealing with the hyperspherical technique. This monograph is aimed at fulfilling this necessity. Besides introducing the hyperspherical variables (which are many-body generalization of ordinary spherical polar coordinates) and hyperspherical harmonics basis for the expansion of a many-body wave function, methods to introduce desired symmetry of the wave function has been discussed. Approximation methods, which simplify the calculations, without loosing sight of the interesting physics sought after, have also been included. Finally, discussion of a number of current topics in physics like

Bose–Einstein condensation, where this technique has been very useful, have been incorporated.

I take this opportunity to thank all my colleagues and friends at the Physics Department of the University of Calcutta for their help and cooperation. I specially thank Prof. A. Raychaudhuri and Prof. P. Sen for their continuous encouragement and support. I also thank all my collaborators, and in particular Prof. B. Chakrabarti, for their help in research. Thanks are due to the Department of Science and Technology, Government of India, for financial assistance for this project (No. HR/UR/08/2011). Last, but not the least, I express my appreciation and thanks to my wife Rita and daughter Arunima, for their understanding and bearing with me, as I could not spare much time for them due to this project.

Catalysed and supported by the Science & Engineering Research Board, Department of Science & Technology under its Utilisation of Scientific Expertise of Retired Scientists Scheme.

Kolkata   Tapan Kumar Das

Contents

About the Author

Dr. **Tapan Kumar Das** completed M.Sc. from Calcutta University and Ph.D. from the University of Pennsylvania (USA) in Theoretical Nuclear Physics under Prof. Abraham Klein and Prof. R.M. Dreizler in 1971. That work involved theoretical foundation of variable moment of inertia models in nuclear theory. He did his Post-doctoral research in nuclear many-body theory at the Technische Universität München (Germany). Then in 1973, he joined the Universidade Federal de Pernambuco (Brazil) as Professor of Physics. He collaborated with other physicists on few nucleon problems, including the proposition of a new tamed nuclear three-body force. Dr. Das returned to India in 1976 and joined the University of Burdwan, where he taught Physics as Lecturer, Reader and finally as Professor up to 1990, while continuing research on few-body systems using hyperspherical harmonics technique. He joined the University of Calcutta as Professor of Physics in 1990 and remained in that position until 2008. After that, he was awarded the UGC (Government of India) Emeritus Fellowship for 2 years and finally the DST (Government of India) Users Book Grant Project for 2 years from 2012. Dr. Das has visited Brazil, France, Russia, Italy and South Africa to do various collaborative research works. His research interests have been use of hyperspherical harmonics to numerous problems of physics and supersymmetric quantum mechanics. His major research contributions are: proposition of a 'tamed three-nucleon force' and study its effect on trinucleon systems using hyperspherical harmonics, development of a many-body technique (correlated potential harmonics expansion method) for the Bose-Einstein condensate, formulation of supersymmetric quantum mechanics to multidimensional systems, etc. He has over 110 publications in reputed international journals to date.

Chapter 1
Introduction

Abstract Need for quantum mechanical solution of few-body to many-body systems in physics is emphasized. Requirement of appropriate symmetry of the wave function under exchange of pairs of identical particles in the system is discussed. Quantum numbers are introduced for classification of states. Good quantum numbers are defined. Hyperspherical harmonics expansion method (HHEM) is an ab initio technique which can be generalized for systems with more than three particles.

In problems of physics, one often needs the quantum mechanical solution of systems containing a few to a moderate number, and eventually to a large number of particles. Examples are as follows:

1. An electron moving in an external field, say an electromagnetic field. This is an example of a single particle moving under the influence of an external potential field.
2. An electron moving in the field produced by a proton, but no externally applied field (the hydrogen atom). This is an example of two particles moving under mutual forces only. In this case, one can separate the two-body motion into a center-of-mass (CM) motion and a relative motion governed by the mutual potential, as in the case of a classical system.
3. Two electrons moving in the field of a nucleus containing two protons and two neutrons, but no externally applied field. This is the simple neutral Helium atom. The nucleus is very strongly bound compared to the Helium atom and its size is very much smaller than the size of the atom. Hence, to a very good approximation, the nucleus can be treated as a single particle and the Helium atom as a three-body system interacting through mutual Coulomb forces.
4. An atom containing N electrons is an example of an (N+1)-body system (assuming the nucleus to be a single particle).
5. A nucleus consists of N neutrons and Z protons. Treating nucleons as 'particles,' the nucleus is an A-body system, where $A = N + Z$. Such a nucleus is usually denoted by the symbol $^A_N X_Z$, where X is its chemical symbol. It is also denoted by the symbol $^A X$, since the chemical symbol X is associated with a unique value of Z. The simplest nucleus, deuteron, is a two-body system consisting of a proton

© Springer India 2016

T.K. Das, *Hyperspherical Harmonics Expansion Techniques,*
Theoretical and Mathematical Physics, DOI 10.1007/978-81-322-2361-0_1

and a neutron. Trinucleon nuclei, $^3_2\text{H}_1$ and $^3_1\text{He}_2$, are examples of nuclear three-body systems. Considering the nucleon as constituted by three interacting quarks, the nucleon itself is a three-body system. The energy and length scales of interest in a particular study determine whether constituents of a 'particle' must be treated individually or as a single entity treated as a particle.

6. The dilute *Bose gas* at a very low temperature is an example of a system containing very large number of mutually interacting bosonic atoms. Below a critical temperature, a macroscopic fraction of such atoms remains in the lowest energy state. The system in such a state is called the *Bose–Einstein condensate*. Although the Bose–Einstein condensate behaves as a single quantum mechanical object, having collective motions, it also exhibits the signatures of a many-body system. In the following chapters, we will see that the quantum mechanical solution of a many-body system is very difficult when the number of particles is greater than three. However, if the system is so dilute that *only two-body correlations* are of relevance, then the many-body Schrödinger equation can be solved to a very good approximation, using the few-body techniques. Such a simplification is valid for the Bose–Einstein condensate.

In most cases, a 'particle' is indeed a bound system of smaller constituents. In all the above examples, we tacitly assume that a 'particle' is a *point particle* having specified mass, spin, isospin, magnetic moment, *etc.* The energy scale under consideration is small enough (it should be small compared to the binding energy against its dissociation into the constituents), so that an aggregate of smaller constituents can be considered as a single entity, to be treated as a 'particle.' If the energy scale is higher (larger than or comparable with the binding energy of the aggregate 'particle'), one has to treat all the constituents as separate particles. For example, in the atomic system (example 4 above), typical energies are of the order of eV, while the binding energy of the nucleus is of the order of MeV. Hence, in this case, the nucleus can be treated as a 'particle.' Moreover, since the mass of the nucleus is much greater than the mass of an electron, the nucleus can be treated as at rest. In what follows, we will consider a physical system consisting of a *fixed number of particles* at *non-relativistic energies*, so that the system is described by the Schrödinger equation.

In many problems of physics, the Hamiltonian is independent of time, at least to a very good approximation. In this case, the total energy of the system (E) is conserved (i.e., remains unchanged in time) and the time dependence of the wave function Ψ separates into a product of a wave function ψ dependent on the space variables of the system and a function of time given by $\exp(-\frac{i}{\hbar}Et)$. The space-dependent wave function ψ satisfies the time-independent Schrödinger equation, which is the eigenvalue equation of the time-independent Hamiltonian with eigenvalue E. Since E is the eigenvalue of a Hermitian operator, it is real. All observable quantities of the system involve bilinear combination of the wave function: the operator corresponding to the observable being sandwiched between Ψ^* and Ψ. In particular, the probability density is the modulus squared of the wave function. Since E is real, the time dependence of all the observables disappears. Such a state of the system is called a *stationary state*, meaning that even though the wave function has a time dependence, the observables of the system (unless they have explicit time dependence) are time independent.

For systems containing two or more particles, there is a possibility that some of the particles are identical. If the identical particles are bosons, i.e., particles of integral intrinsic spin $(0, 1, 2, \ldots$ in units of $\hbar)$, the wave function must be symmetric under the exchange of every pair of such particles. On the other hand, if the particles are fermions, i.e., particles of half-integral spin $(\frac{1}{2}, \frac{3}{2}, \frac{5}{2}, \ldots)$, the wave function must be antisymmetric under exchange of every pair of such particles. Exchanging a pair of particles means interchanging all the dynamical variables *viz.*, position, spin, isospin, etc. Thus new degrees of freedom, like the spin and isospin, also come into the picture, even if the Hamiltonian is independent of them. A system may contain a number of identical bosons and a number of identical fermions. Then the total many-body wave function should be symmetric under exchange of any pair of identical bosons and antisymmetric under the exchange of any pair of identical fermions. However, no symmetry is required under the exchange of nonidentical particles.

For the complete classification of the states, we have also to specify the quantum numbers of the system. For simplicity, let us ignore isospin for this discussion. The total orbital angular momentum of a system of A particles is the vector sum of orbital angular momenta of all the particles: $\vec{L} = \vec{l}_1 + \vec{l}_2 + \cdots + \vec{l}_A$. The vector coupling of the angular momenta will involve $(A - 2)$ intermediate angular momenta. Likewise the total spin of the system is $\vec{S} = \vec{s}_1 + \vec{s}_2 + \cdots + \vec{s}_A$. Total angular momentum of the system is $\vec{J} = \vec{L} + \vec{S}$. There are five quantum numbers associated with the ith particle: n_i associated with the radial motion, l_i, m_{l_i} associated with the angular degrees of freedom, and s_i, m_{s_i} associated with the spin degrees of freedom. However, particles have a fixed spin value (for example, electrons have spin $\frac{1}{2}$) and so only four quantum numbers are necessary. Hence, a total of $4A$ quantum numbers are needed for a complete classification. Not all of these are *good quantum numbers*. A quantum number is said to be 'good' when the corresponding operator commutes with the Hamiltonian. A quantum number which is not good is not conserved. i.e., an eigen state of the Hamiltonian will, in general, be a mixture of states with different values of such nonconserved quantum numbers. For a system of particles with mutual two-body spin-independent central interactions only, with no external forces, L and S are good and one can use the LS-coupling scheme. In this scheme, individual orbital and spin angular momenta are vector coupled to total orbital $(\vec{L} = \vec{l}_1 + \vec{l}_2 + \cdots + \vec{l}_A)$ and total spin $(\vec{S} = \vec{s}_1 + \vec{s}_2 + \cdots + \vec{s}_A)$ angular momenta. The resultant total orbital and total spin angular momenta are then vector coupled to the *total angular momentum* $(\vec{J} = \vec{L} + \vec{S})$. On the other hand, for spin-dependent mutual forces (for example, with spin-orbit terms), the jj-coupling scheme should be used, with the total angular momentum of the ith particle, $\vec{j}_i = \vec{l}_i + \vec{s}_i$ and the total angular momentum of the system $\vec{J} = \vec{j}_1 + \vec{j}_2 + \cdots + \vec{j}_A$.

It is obvious that an *ab initio* solution of the Schrödinger equation for a many-body system is the most desirable one. However, the complexity of the space part alone increases gallopingly as the number of particles increases from two to three and more. Moreover, the couplings of orbital and spin (as also isospin, where it is relevant) angular momenta become very complicated as A increases. In the following chapters, we will discuss the hyperspherical harmonics expansion method (HHEM) for gradually more complicated systems as A increases from 3. Fortunately, a large number of important physical systems involve up to three particles, which can be solved by this technique using the present-day numerical methods in modern computers (see Chaps. 3, 5 and 6). Systems containing four or five particles have been solved by such techniques with some approximations (see Chap. 4). The method can be utilized in some bigger systems, where bound clusters of a few particles can be regarded as a 'single particle.' Examples are alpha particle model of light $A = 4n$ nuclei with $n = 3, 4, \ldots$ (see Chap. 4). Monte Carlo technique in the few-body hyperspherical harmonics method has been successfully applied to systems containing up to 100 particles. A large number of particles can also be handled approximately, using a subset called *potential harmonics* of the full hyperspherical harmonics set (see Chap. 7) for very dilute systems, like the Bose–Einstein condensates (BEC), as mentioned in example 6 above. This will be discussed in Chap. 8.

Chapter 2
Systems of One or More Particles

Abstract A particle moving in an external field is an example of one-body system. The Schrödinger equation reduces to a single differential equation for central potential. The relative motion of a two-body system with mutual potential reduces to one-body Schrödinger equation. Symmetry of the wave function for identical particles is discussed. In this connection, spin and isospin and wave functions involving them are introduced. Next many-body wave equation is written down and the need for approximations and models stressed. Mean-field approximation and independent particle model are introduced.

2.1 One-Body System: A Particle in a Potential Field

We first consider the simplest system consisting of a single particle of mass m moving in an external time-independent potential $V(\vec{r})$. In the following, we review this topic, which can be found in standard texts in quantum mechanics, for example references [1–6]. The time-independent Schrödinger equation is

$$\left[-\frac{\hbar^2}{2m} \nabla^2 + V(\vec{r}) \right] \psi(\vec{r}) = E\psi(\vec{r}), \qquad (2.1)$$

where $\vec{r}$ is the position vector of the particle of mass m and the Laplacian is

$$\nabla^2 = \frac{\partial^2}{\partial x^2} + \frac{\partial^2}{\partial y^2} + \frac{\partial^2}{\partial z^2}$$

$$= \frac{1}{r^2}\frac{\partial}{\partial r}\left(r^2 \frac{\partial}{\partial r} \right) + \frac{1}{r^2 \sin\theta}\frac{\partial}{\partial \theta}\left(\sin\theta \frac{\partial}{\partial \theta} \right) + \frac{1}{r^2 \sin^2\theta}\frac{\partial^2}{\partial \phi^2}, \qquad (2.2)$$

(x, y, z) and (r, θ, ϕ) being the Cartesian and spherical polar coordinates of $\vec{r}$. The Laplacian can also be expressed in terms of the orbital angular momentum operator $\hat{L}^2$

$$\nabla^2 = \frac{1}{r^2}\frac{\partial}{\partial r}\left(r^2 \frac{\partial}{\partial r} \right) - \frac{\hat{L}^2}{r^2 \hbar^2}, \qquad (2.3)$$

© Springer India 2016

T.K. Das, *Hyperspherical Harmonics Expansion Techniques*,
Theoretical and Mathematical Physics, DOI 10.1007/978-81-322-2361-0_2

since $\hat{L}^2$ is given in spherical polar coordinates as

$$\hat{L}^2 = -\hbar^2 \left[\frac{1}{\sin\theta} \frac{\partial}{\partial\theta} \left(\sin\theta \frac{\partial}{\partial\theta} \right) + \frac{1}{\sin^2\theta} \frac{\partial^2}{\partial\phi^2} \right]. \tag{2.4}$$

The operator $\hat{L}^2$ satisfies an eigenvalue equation

$$\hat{L}^2 Y_{lm}(\theta, \phi) = \hbar^2 l(l+1) Y_{lm}(\theta, \phi), \tag{2.5}$$

with eigenvalue $\hbar^2 l(l+1)$ and the spherical harmonics $Y_{lm}(\theta, \phi)$ as the corresponding eigenfunction. Note that each component of $\hat{\vec{L}}$ commutes with $\hat{L}^2$, but components of $\hat{\vec{L}}$ do not commute among themselves in pairs. Hence, *only one component* of $\hat{\vec{L}}$ and $\hat{L}^2$ can simultaneously be specified. Conventionally, this component is chosen as the z-component $\hat{L}_z$, with eigenvalue $m\hbar$. $Y_{lm}(\theta, \phi)$ is the simultaneous eigenfunction of $\hat{L}^2$ and $\hat{L}_z$.

If the potential is spherically symmetric, i.e., independent of the direction (θ, ϕ), so that $V(\vec{r}) = V(r)$ (such a potential is called *central*, since then the force is always directed along the line joining the particle with the center), then the orbital angular momentum (l) and its projection (m) are good quantum numbers (since the Hamiltonian commutes with both $\hat{L}^2$ and $\hat{L}_z$) and the wave function has the form

$$\psi_{nlm}(\vec{r}) = \frac{R_{nl}(r)}{r} Y_{lm}(\theta, \phi). \tag{2.6}$$

The factor $\frac{1}{r}$ is included to remove the first derivative with respect to r. The radial Schrödinger equation satisfied by $R_{nl}(r)$ becomes

$$\left[-\frac{\hbar^2}{2m} \frac{d^2}{dr^2} + \frac{\hbar^2 l(l+1)}{2mr^2} + V(r) - E \right] R_{nl}(r) = 0. \tag{2.7}$$

This second-order ordinary differential equation can be solved, subject to appropriate boundary conditions to get the energy eigenvalue E and the radial wave function $R_{nl}(r)$. For a bound state, the probability of finding the particle must be finite only within a finite region of space and vanish at great distances from the center. Hence, $R_{nl}(r)$ must be square integrable and vanish for $r \to \infty$. From Eq. (2.7), we see that this requirement is satisfied if $E < V(\infty)$. Moreover, since ψ should be finite everywhere, Eq. (2.6) shows that $R_{nl}(r)$ must also vanish at the origin. Imposition of these boundary conditions at the origin and at $r \to \infty$ makes energy eigenvalue (E) *discrete* and less than $V(\infty)$ for the bound state. The *quantum number n* is usually associated with numbering the discrete energy eigenvalues consecutively with increasing energy, for a particular l. Taking $n = 0$ for the lowest energy state for a particular value of l, n is the number of nodes in the radial wave function $R_{nl}(r)$.

For the unbound scattering state, probability of finding the particle at a large distance is nonvanishing. Hence, $E > V(\infty)$ and energy eigenvalues form a *continuum* [7]. In this case, the discrete quantum number n is replaced by the energy eigenvalue E. The boundary conditions on the radial wave function are $R_{E,l}(r)$ vanishes at the origin, while it oscillates with a finite amplitude as $r \rightarrow \infty$.

If the potential has a simple form, like an infinite square well, a harmonic oscillator, or a Coulomb potential, the radial equation can be solved analytically, although for most other potentials, analytic solutions are not possible. What is the inherent reason that only some potentials admit exact analytic solutions? This can be understood from the concept of supersymmetric quantum mechanics. Using the technique of supersymmetric quantum mechanics [8], one can obtain a supersymmetric partner potential for any given potential. Then it can be shown that exact algebraic solutions can be obtained for a potential, if the potential and its supersymmetric partner have the same mathematical shape, but involves different parameters. Such a potential is called a *shape invariant potential*. It has been shown that only a few potentials belong to this category, the infinite square well, the harmonic oscillator, and the Coulomb potentials being some of them [8].

Unfortunately, most of the potentials appearing in physical problems do not permit any analytical solution. In such cases, we need to solve the Schrödinger equation numerically. Usually, the second-order differential equation (2.7) is written as a system of two coupled first-order differential equations, which are then solved by a standard technique, e.g., the Runga–Kutta algorithm [9]. Choosing an initial trial energy, the integration is done in two parts: outward from the origin to a suitably chosen match radius (r_M) and inward from a large enough value of r to r_M. Next the trial energy is changed until the log derivatives at $r = r_M$ obtained from the inward and outward integrations match (usually done by a root-finding algorithm [9]). This gives the eigen energy. The wave function obtained at this energy is then normalized to get the final normalized wave function. The method will be discussed in detail in Chap. 10.

2.2 Two-Body System with Mutual Interaction

We now consider the next more complex system, *viz.*, the two-body system interacting through mutual forces. In addition to solving the two-body Schrödinger equation, we have to worry about the symmetry of the wave function, if the particles are identical [1]. This brings the spin of the particle into consideration, even if the potential is spin independent. If, in addition, the mutual force depends on the spins of the particles, additional terms in the Hamiltonian coming from the spins are to be considered. Thus we see that the system already starts to be complicated, even when it contains only two particles. First, we discuss how the two-body Schrödinger equation in space variables can be solved in a suitable way.

2.2.1 Two Distinct Particles

For a system consisting of two particles of masses m_1 and m_2 with position vectors $\vec{r}_1$ and $\vec{r}_2$ and interacting through a mutual time-independent potential $V(\vec{r}_1 - \vec{r}_2)$, the Schrödinger equation is (for the moment we ignore spin variables)

$$\left[-\frac{\hbar^2}{2m_1}\nabla_1^2 - \frac{\hbar^2}{2m_2}\nabla_2^2 + V(\vec{r}_1 - \vec{r}_2) - E_T \right] \Psi(\vec{r}_1, \vec{r}_2) = 0, \qquad (2.8)$$

E_T is the total energy of the system and

$$\begin{aligned}
\nabla_i^2 &= \frac{\partial^2}{\partial x_i^2} + \frac{\partial^2}{\partial y_i^2} + \frac{\partial^2}{\partial z_i^2} \\
&= \frac{1}{r_i^2}\frac{\partial}{\partial r_i}\left(r_i^2 \frac{\partial}{\partial r_i} \right) + \frac{1}{r_i^2 \sin\theta_i}\frac{\partial}{\partial\theta_i}\left(\sin\theta_i \frac{\partial}{\partial\theta_i} \right) + \frac{1}{r_i^2 \sin^2\theta_i}\frac{\partial^2}{\partial\phi_i^2}. \quad (2.9)
\end{aligned}$$

where (x_i, y_i, z_i) are the Cartesian coordinates and (r_i, θ_i, ϕ_i) are the spherical polar coordinates of $\vec{r}_i$ ($i = 1, 2$). In order to solve this equation, one can easily separate the relative motion from the center-of-mass motion, by introducing the relative vector [7]

$$\vec{r} = \vec{r}_1 - \vec{r}_2 \qquad (2.10)$$

and the center-of-mass vector

$$\vec{R} = \frac{m_1\vec{r}_1 + m_2\vec{r}_2}{m_1 + m_2}. \qquad (2.11)$$

By a straight forward evaluation of the partial derivatives, we can verify that

$$\frac{1}{m_1}\nabla_1^2 + \frac{1}{m_2}\nabla_2^2 = \frac{1}{\mu}\nabla_r^2 + \frac{1}{M}\nabla_R^2, \qquad (2.12)$$

where ∇_r^2 and ∇_R^2 are the Laplacians with respect to $\vec{r}$ and $\vec{R}$, respectively, $M = m_1 + m_2$ is the total mass and

$$\mu = \frac{m_1 m_2}{m_1 + m_2} \qquad (2.13)$$

is the *reduced mass* of the system. Since the Laplacian of the two-body system separates into the sum of Laplacians of the relative and the center-of-mass coordinates and the potential is a function of $\vec{r}$ only, the eigenfunction Ψ [expressed as a function of $\vec{r}$ and $\vec{R}$, using Eqs. (2.10) and (2.11)] becomes separable into a product of the relative and the center-of-mass wave functions

$$\Psi(\vec{r}, \vec{R}) = \psi(\vec{r})\Phi(\vec{R}). \qquad (2.14)$$

Substituting Eq. (2.12) in Eq. (2.8) and dividing throughout by Ψ, we have

$$\frac{1}{\psi(\vec{r})}\left[-\frac{\hbar^2}{2\mu}\nabla_r^2 + V(\vec{r})\right]\psi(\vec{r}) = \frac{1}{\Phi(\vec{R})}\left[\frac{\hbar^2}{2M}\nabla_R^2\right]\Phi(\vec{R}) + E_T. \qquad (2.15)$$

We note that the left side is a function of $\vec{r}$ only, while the right side is a function of $\vec{R}$ only. This can be true for arbitrary values of $\vec{r}$ and $\vec{R}$, only if each side is a constant, independent of $\vec{r}$ and $\vec{R}$. Calling this separation constant E, we have

$$\left[-\frac{\hbar^2}{2\mu}\nabla_r^2 + V(\vec{r})\right]\psi(\vec{r}) = E\psi(\vec{r}) \qquad (2.16)$$

$$\left[-\frac{\hbar^2}{2M}\nabla_R^2\right]\Phi(\vec{R}) = E_{CM}\Phi(\vec{R}), \qquad (2.17)$$

where $E_{CM} = E_T - E$ is the energy of the center-of-mass motion, E is the energy of the relative motion, and E_T is the total energy. Equation (2.17) shows that the center-of-mass of the system moves like a *free particle* of mass M (sum of the individual masses) and momentum $\vec{P} = \hbar\vec{K}$

$$\Phi(\vec{R}) = Ce^{i\vec{K}.\vec{R}}, \qquad (2.18)$$

where C is a normalization constant and $E_{CM} = \frac{\hbar^2 K^2}{2M}$. This is expected, since there is no *external force* and the entire system of two particles moves like a free body as a single entity. Since only the relative motion of the two particles under their mutual force is of physical interest, we need to consider only Eq. (2.16). This is the same as the one-body Schrödinger equation (2.1). Thus a two-body system under mutual force only is equivalent to the motion of a single *fictitious particle* of mass μ moving in *a potential field* $V(\vec{r})$. This equation can be solved as discussed in the last section.

2.2.2 Two Identical Particles: Symmetry of Wave Function

As mentioned at the beginning of this section, we have to worry about the symmetry of the wave function, if the particles are identical (hence $m_1 = m_2$). The two-body wave function should be symmetric for two identical bosons (integral spin particles) and antisymmetric for two identical fermions (half-integral spin particles).

We first discuss the symmetry of the space part of the wave function, ignoring the spin degrees of freedom for the time being. The symmetric (upper sign) and antisymmetric (lower sign) wave functions under pair exchange operator P_{12} satisfy

$$P_{12}\Psi(\vec{r}_1, \vec{r}_2) = \Psi(\vec{r}_2, \vec{r}_1) = \pm\Psi(\vec{r}_1, \vec{r}_2). \qquad (2.19)$$

Now, Eqs. (2.10) and (2.11) show that under this exchange $\vec{r}$ becomes $-\vec{r}$, while $\vec{R}$ remains unchanged. Hence, the center-of-mass motion remains unaffected while the relative wave function $\psi(\vec{r})$ becomes $\psi(-\vec{r})$. This corresponds to the parity operation, $\vec{r} \rightarrow -\vec{r} = (r, \pi - \theta, \pi + \phi)$, and the spherical harmonics in Eq. (2.6) becomes

$$Y_{lm}(\pi - \theta, \pi + \phi) = (-1)^l Y_{lm}(\theta, \phi). \tag{2.20}$$

Hence, Eq. (2.19) will be satisfied only if l is even for upper sign (symmetric wave function) and odd for lower sign (antisymmetric wave function). In the following, we will see how the full wave function (including spin wave function) should be properly symmetrized.

2.2.3 Inclusion of Spin Degrees of Freedom

Next we consider a two-body system in which spin degrees of freedom come into play. In the most general case, the forces may depend on spin, position, and orbital angular momentum. The total spin S (where $\vec{S} = \vec{s}_1 + \vec{s}_2$, $\vec{s}_i$ being the spin operator of the ith particle, $i = 1, 2$) and its projection M_S, as well as the orbital angular momentum l (where $\vec{l} = \vec{l}_1 + \vec{l}_2$, $\vec{l}_i$ being the orbital angular momentum of the ith particle), and its projection m_l are good quantum numbers, if either the total mutual force is spin independent or if it contains $\vec{s}_1 \cdot \vec{s}_2$ only. However, if there is spin-orbit force $(\vec{l}.\vec{S})$ the good quantum numbers are l, S, J, M_J where J and M_J are the eigenvalues corresponding to the total angular momentum $(\vec{J} = \vec{l} + \vec{S})$ and its projection. For tensor force [given by $3(\vec{\sigma}_1.\hat{r})(\vec{\sigma}_2.\hat{r}) - (\vec{\sigma}_1.\vec{\sigma}_2)$, where $\vec{\sigma}_i$ is the Pauli spin operator for the ith particle], there can be l mixing.

First, we consider two distinct particles for the simple case where l, S, m_l, M_S are good. Typical examples are the simple model of the hydrogen atom, where the interaction between the proton and the electron is spin independent or a simple model of the deuteron nucleus, where the interaction between the proton and the neutron depends on S (ignoring the tensor force). For the simple model of the hydrogen atom, since spin does not appear at all, the spin wave function is either singlet ($S = 0$) or triplet ($S = 1$) and is commonly suppressed. The most important part of nuclear force is spin dependent (contains $\vec{s}_1.\vec{s}_2$), the force in the triplet state being more strongly attractive than that in the singlet state. In the simple model of deuteron (with proton and neutron treated as distinct particles), the singlet ($S = 0$) and triplet ($S = 1$) states are independently calculated (temporarily suppressing spin variables) with singlet and triplet potentials, respectively.

The total wave function now becomes a product of space and spin wave functions

$$\Psi^{Total}(1, 2) \equiv \Psi^{Total}(\vec{r}_1, \vec{s}_1, \vec{r}_2, \vec{s}_2) = \Psi(\vec{r}_1, \vec{r}_2)\chi(\vec{s}_1, \vec{s}_2), \tag{2.21}$$

where $\vec{s}_i$ is the spin variable, given by the spin and its projection operators, $\{\hat{s}_i, \hat{s}_{iz}\}$, of the ith particle ($i = 1, 2$). If the particles are distinct, there is no requirement of symmetry of the total wave function. If the interaction is spin independent (as in the case of simple hydrogen atom), $\chi(\vec{s}_1, \vec{s}_2)$ can take any of the possible S, M_S values, while $\Psi(\vec{r}_1, \vec{r}_2)$ remains the same. If there is spin-dependent force (as in the simple deuteron model), for each possible $\chi(\vec{s}_1, \vec{s}_2)$, the space wave function $\Psi(\vec{r}_1, \vec{r}_2)$ will be different, depending on the potential, which depends on $\vec{s}_1.\vec{s}_2$.

Next consider two identical particles. Since the *total wave function* has to be symmetric for identical bosons and antisymmetric for identical fermions, the product $\Psi \, \chi$ should have the appropriate symmetry. Thus for bosons, both space and spin wave functions must be symmetric or both must be antisymmetric. For fermions, if one is symmetric, the other must be antisymmetric. We already saw that Ψ is symmetric or antisymmetric for even or odd l, respectively. The spin wave function χ is symmetric or antisymmetric for $S = 1$ or $S = 0$, respectively. If l, S, m_l, M_S are good (no mixing of l and S), we simply have to combine appropriate symmetry of Ψ and χ in Eq. (2.21). If there is mixing of l or S, the right side of Eq. (2.21) should be a sum of appropriate combinations of Ψ and χ, corresponding to possible (l, S) values.

Two-Body Spin Wave Function

Denoting the spin state of the ith particle by $|s_i, m_{s_i}\rangle$ (where m_{s_i} is the eigenvalue of $\hat{s}_{iz}$), the two-body spin wave function becomes

$$\chi(\vec{s}_1, \vec{s}_2) \equiv |s_1, s_2, S, M_S\rangle = \sum_{m_{s_1}, m_{s_2}} \langle s_1, m_{s_1}, s_2, m_{s_2} | S, M_S\rangle |s_1, m_{s_1}\rangle |s_2, m_{s_2}\rangle,$$

$$(2.22)$$

where $\langle s_1, m_{s_1}, s_2, m_{s_2} | S, M_S\rangle$ is a Clebsch–Gordan (CG) coefficient. The spin wave function is represented by the ket vector. S and M_S are, respectively, the total spin and its projection

$$\vec{S} = \vec{s}_1 + \vec{s}_2,$$
$$M_S = m_{s_1} + m_{s_2}. \qquad (2.23)$$

By angular momentum selection rule $|s_1 - s_2| \le S \le (s_1 + s_2)$, i.e., S can take the values $|s_1 - s_2|, |s_1 - s_2| + 1, |s_1 - s_2| + 2, \ldots, (s_1 + s_2)$. It can easily be seen that the state with the maximum spin $S = s_1 + s_2$ will be symmetric, the state with next lower S will be antisymmetric, the next lower one symmetric, and so on.

2.2.4 Introduction of Isospin Degrees of Freedom for Nucleons

The nucleons, *viz.*, the proton and the neutron, have nearly the same mass and both are spin $\frac{1}{2}$ particles. It is known experimentally that the nuclear parts of their interactions are the same, at least at low energies. So they can be treated as two states of a single particle called *nucleon*. A fictitious 'spin,' in analogy with the common spin, is associated with the nucleon and called *isospin*, such that the proton and the neutron are two states with a different 'isospin projection' of the *nucleon*. Since there are only two projection states of the nucleon, its isospin is $t = \frac{1}{2}$. The projections (called t_{i_3} instead of t_{i_z}, since $\vec{t}_i$ is not really a vector in three-dimensional coordinate space) of proton and the neutron are $\frac{1}{2}$ and $-\frac{1}{2}$, respectively. Note that this is a purely mathematical construction for convenience and the *isospin* is not a physical quantity. If $\vec{t}_i$ is the isospin of the ith particle, the total isospin of two nucleons is $\vec{T} = \vec{t}_1 + \vec{t}_2$. Then in exact analogy with Eq. (2.22), one can construct isospin state $|t_1, t_2, T, M_T\rangle$ of two nucleons. Clearly, T can take two values, *viz.* 0 and 1, which are, respectively, antisymmetric and symmetric under P_{12} in the isospin space. With the introduction of this isospin degree of freedom, the total wave function becomes

$$\Psi^{Total}(1, 2) \equiv \Psi^{Total}(\vec{r}_1, \vec{s}_1, \vec{t}_1; \vec{r}_2, \vec{s}_2, \vec{t}_2) = \Psi(\vec{r}_1, \vec{r}_2)\chi(\vec{s}_1, \vec{s}_2)\tau(\vec{t}_1, \vec{t}_2), \quad (2.24)$$

where $\tau(\vec{t}_1, \vec{t}_2)$ is the two-body isospin wave function. The total wave function must be antisymmetric under P_{12}, i.e., out of the space, spin, and isospin wave functions, either all three can be antisymmetric, or only one antisymmetric and the other two symmetric. Thus for two nucleons, $T = 1$ states can be spin-triplet odd l (abbreviated as triplet-odd) or spin-singlet even l (singlet-even). Similarly, $T = 0$ states can be triplet-even or singlet-odd. Note that with the introduction of isospin, the deuteron is a two-body system of identical particles. However, without the isospin, it is a system of two distinct particles. Both the descriptions are valid, although imposition of symmetry actually reduces the number of possibilities and thus simplifies the problem.

2.3   System of Several Particles

The situation becomes already complicated, when three mutually interacting particles are involved, since it will involve two relative vectors, after removal of the CM motion. Clearly, the situation worsens rapidly as the number of interacting particles in the system increases. In the next chapter, we will discuss the three-body system in detail. In this section, we discuss some general features of the A-body system.

The Schrödinger equation for a system of A particles of mass m_i ($i = 1, A$), interacting through pair-wise mutual forces, is

$$\left[-\sum_{i=1}^{A} \frac{\hbar^2}{2m_i}\nabla_i^2 + \sum_{i<j=2}^{A} V(\vec{r}_i - \vec{r}_j) \right] \Psi(\vec{r}_1, \vec{r}_2, \dots, \vec{r}_A) = E\Psi(\vec{r}_1, \vec{r}_2, \dots, \vec{r}_A).$$

$$(2.25)$$

The restriction $i < j = 2, A$ in Eq. (2.25) is needed to avoid double counting of pair-wise interactions. In this case also, we can separate the center-of-mass (CM) motion in terms of the CM vector $\vec{R}$. In addition, there will be $N = (A - 1)$ relative vectors to describe the space part of the relative motion. For a mutually interacting system without any externally applied field, the CM motion is not important, and only the relative motion has to be studied. This involves $(3A - 3)$ degrees of freedom for the space part alone. Associated algebraic procedure is very involved, as we will see in the following chapters. Imposition of symmetry is also very complicated. For $A \geq 3$, there will be states of mixed symmetry, in addition to states, which are totally symmetric or totally antisymmetric under pair exchanges. Introduction of spin of the particles (and isospin, if it is relevant) will further complicate the form of the total wave function. For a system of identical bosons or fermions, the total wave function with the appropriate symmetry will be obtained as a sum of products of space and spin (and also isospin, where the latter is relevant) wave functions of conjugate symmetry, such that each product has the desired symmetry. Clearly, this is going to be a nontrivial exercise and will be dealt in the following chapters.

2.3.1 Independent Particle Model: Mean-Field Description

However, the situation simplifies immensely, if the total interaction of a system of identical particles (with $m_i = m, i = 1, A$) is separable and $\sum_{i<j=2}^{A} V(\vec{r}_i - \vec{r}_j)$ can be replaced by $\sum_{i=1}^{A} \overline{V}(\vec{r}_i)$, such that each individual particle is subjected to the *same* potential field $\overline{V}(\vec{r})$. Then Eq. (2.25) is replaced by

$$\left[-\sum_{i=1}^{A} \frac{\hbar^2}{2m} \nabla_i^2 + \sum_{i=1}^{A} \overline{V}(\vec{r}_i) \right] \Psi(\vec{r}_1, \vec{r}_2, \ldots, \vec{r}_A) = E \Psi(\vec{r}_1, \vec{r}_2, \ldots, \vec{r}_A). \quad (2.26)$$

In this case, the many-body wave function is separable into a product of single-particle (s.p.) wave functions

$$\Psi(\vec{r}_1, \vec{r}_2, \ldots, \vec{r}_A) = \prod_{i=1}^{A} \psi_{\alpha_i}(\vec{r}_i), \quad (2.27)$$

with

$$E = \sum_{i=1}^{A} \epsilon_{\alpha_i}, \quad (2.28)$$

where $\psi_{\alpha_i}(\vec{r}_i)$ satisfies a single particle Schrödinger equation

$$\left[-\frac{\hbar^2}{2m} \nabla_i^2 + \overline{V}(\vec{r}_i) \right] \psi_{\alpha_i}(\vec{r}_i) = \epsilon_{\alpha_i} \psi_{\alpha_i}(\vec{r}_i). \quad (2.29)$$

Here α_i represents an abbreviated form of s.p. quantum numbers (a set of three quantum numbers for the space wave function) associated with a particular s.p. state $|\alpha_i\rangle$ of the Hamiltonian $\hat{h}_i = -\frac{\hbar^2}{2m}\nabla_i^2 + \overline{V}(\vec{r}_i)$. The full Hamiltonian of the system is $\hat{H} = \sum_i \hat{h}_i$. Since $\hat{h}_i$ is the same for all i, the subscript i can be dropped in Eq. (2.29). Thus the many-body problem reduces immediately to the solution of a one-body problem, which is easy to solve and has already been discussed.

It may appear that the satisfaction of the condition $\sum_{i<j=2}^A V(\vec{r}_i - \vec{r}_j) = \sum_{i=1}^A \overline{V}(\vec{r}_i)$ is extremely fortuitous and very unlikely in a real physical situation. Indeed, we may never be so lucky as to have this condition exactly satisfied. However, in a real situation like a nucleus (with many nucleons) or an atom (with many electrons), it may not be very far from truth. In such a case, each one of the particles (nucleons or electrons) moves in the *same* common field produced by all the remaining particles in the system. Hence, to a good approximation, each particle moves in the *same mean field* $\overline{V}(\vec{r})$, produced as a result of interactions and motions of all the remaining particles. This physical scenario may be understood by the following crude analogy. Consider a merry-go-round, in which each individual can go in completely random and fast orbital motion. Then a particular individual will not 'see' any other particular individual, but only a cloud ('field') around himself, due to fast relative motions of all others. However, every other individual will 'see' the *same* cloud. Thus, the sum total of all the individual interactions will effectively be seen as each individual experiencing the same common field. This is the physical basis of the *mean-field theory* (MFT). This leads to the highly successful *shell model* in a nucleus or in an atom. The difference $\sum_{i<j=2}^A V(\vec{r}_i - \vec{r}_j) - \sum_{i=1}^A \overline{V}(\vec{r}_i)$, called *residual interaction*, can be treated as a perturbation. In the MFT, the mean field $\overline{V}(\vec{r})$ can be calculated using suitably chosen trial single-particle wave function $\psi_{\alpha_i}(\vec{r}_i)$. This is then substituted in Eq. (2.29) and a fresh set of $\psi_{\alpha_i}(\vec{r}_i)$ recalculated. This process is repeated until convergence is achieved.

If the particles in the system (a nucleus or an atom in the above example) are identical, we have to symmetrize the wave function appropriately. Since the particles are identical, any permutation of the particle indices leaves the Hamiltonian invariant. Thus the system obeys permutation symmetry and energy eigenvalues are degenerate under permutation operations. We can see this from Eqs. (2.27) and (2.28). Any permutation of the particle indices produces another wave function which is different and orthogonal to the original one, but corresponds to the same energy. Hence, any linear combination of these degenerate eigenfunctions is also a solution belonging to the same energy. The symmetrization postulate demands that the physical wave function be totally symmetric or totally antisymmetric under any pair exchange, for a system of identical bosons or identical fermions, respectively. Hence, totally symmetric or totally antisymmetric linear combinations of Eq. (2.27) should be chosen for a boson system or a fermion system, respectively. Now, any permutation $(\hat{P})$ can be obtained as a set of n_P successive pair exchanges. Hence, the symmetric wave function is

$$\Psi^{(symmetric)}(\vec{r}_1, \vec{r}_2, \ldots, \vec{r}_A) = \frac{1}{\sqrt{A!}} \sum_{\hat{P}} \hat{P} \prod_{i=1}^{A} \psi_{\alpha_i}(\vec{r}_i), \qquad (2.30)$$

and the antisymmetric wave function is

$$\Psi^{(antisymmetric)}(\vec{r}_1, \vec{r}_2, \ldots, \vec{r}_A) = \frac{1}{\sqrt{A!}} \sum_{\hat{P}} (-1)^{n_P} \hat{P} \prod_{i=1}^{A} \psi_{\alpha_i}(\vec{r}_i). \qquad (2.31)$$

The factor $\frac{1}{\sqrt{A!}}$ in front is the normalization constant, since each s.p. wave function $\psi_{\alpha_i}(\vec{r}_i)$ [as also $\Psi(\vec{r}_1, \vec{r}_2, \ldots, \vec{r}_A)$] is normalized and there are $A!$ permutations. As spin (and isospin for nucleons) degrees are relevant for identical particles, the abbreviated quantum number α_i will now include spin (and isospin for nucleons) and their projections. Including the spin degrees of freedom, the variable $\vec{r}_i$ is replaced by the abbreviated notation i (for the ith particle), which stands for $\{\vec{r}_i, \vec{s}_i\}$. With the addition of isospin, i represents $\{\vec{r}_i, \vec{s}_i, \vec{t}_i\}$. With this notation, the antisymmetric wave function, Eq. (2.31), is just the determinantal wave function

$$\Psi^{(antisymmetric)}(1, 2, \ldots, A) = \frac{1}{\sqrt{A!}} \begin{vmatrix} \psi_{\alpha_1}(1) & \psi_{\alpha_1}(2) & \ldots & \psi_{\alpha_1}(A) \\ \psi_{\alpha_2}(1) & \psi_{\alpha_2}(2) & \ldots & \psi_{\alpha_2}(A) \\ \ldots & \ldots & \ldots\ldots \\ \ldots & \ldots & \ldots\ldots \\ \psi_{\alpha_A}(1) & \psi_{\alpha_A}(2) & \ldots & \psi_{\alpha_A}(A) \end{vmatrix}. \qquad (2.32)$$

This determinant is called the *Slater determinant*. Any pair exchange of two particles corresponds to exchanging two columns of this determinant, which results in the same determinant with a negative sign. Thus, it clearly shows that this wave function is antisymmetric under any pair exchange.

2.3.2 Many-Body Description

The discussion in the previous subsection considers the total wave function as a product wave function. This is an independent particle model of the system and is exact only if the total Hamiltonian is separable in the particle indices. For an exact treatment of a realistic case, Eq. (2.25) must be solved. As we explained earlier, the treatment of the space part becomes very complicated as the number of particles increases. In addition, imposition of the required symmetry of the total wave function becomes quite involved. This is because unlike the simple product wave function where only a particular linear combination of the s.p. wave functions gives the desired symmetry, one has to sum combinations of space wave function of a particular (including mixed) symmetry with spin wave function (and isospin wave function, where it is relevant) of the conjugate symmetry. We will see how this can be done for the trinucleon system in Chap. 5.

References

1. Sakurai, J.J.: Modern Quantum Mechanics, 2nd edn. Addison-Wesley, Delhi (2000) (Indian reprint)
2. Greiner, W.: Quantum Mechanics: An Introduction. Springer, New Delhi, 2004; Quantum Mechanics: Special Chapters. Springer, Berlin (1998)
3. Shankar, R.: Principles of Quantum Mechanics, 2nd edn. Springer, New Delhi (2008) (Indian Reprint)
4. Goswami, A.: Quantum Mechanics. Overseas Press, New Delhi (2009)
5. Scheck, F.: Quantum Mechanics. Springer, Bernin (2007)
6. Reed, B.C.: Quantum Mechanics. Jones and Bartlett Publishers, New Delhi (2010)
7. Schiff, L.I.: Quantum Mechanics, 3rd edn. McGraw-Hill Book Company Inc., Singapore (1968)
8. Cooper, F., Khare, A., Sukhatme, U.: Phys. Rep. **251**, 267 (1995)
9. Press, W.H., Teukolsky, S.A., Vetterling, W.T., Flannery, B.P.: Numerical Recipes in FORTRAN: The Art of Scientific Computing, 2nd edn. Cambridge University Press, Cambridge (1992)

Chapter 3
Three-Body System

Abstract Hyperspherical harmonics (HH) expansion method is introduced for the three-body system. Jacobi coordinates are defined, in terms of which the center of mass motion separates for a mutually interacting system. Hyperspherical coordinates and hyperangular momentum are introduced. Analytical expression for the eigen-function (called HH) of the latter is derived. Expanding the relative wave function in the complete basis of HH and substituting in the Schrödinger equation, a system of coupled differential equation is derived. Symmetrization of the HH basis using the Raynal–Revai coefficients (RRC) is discussed. Calculation of the potential matrix elements (PME) is facilitated by multipolar expansion of the potential in the HH basis. Then PME becomes a sum of products of potential multipoles and geometrical structure coefficients (GSC). An elegant method for calculation of GSC is developed using the completeness property of the Jacobi polynomials. An explicit expression is obtained for central potentials.

In this chapter we consider the three-body system, which comes next to the two-body system, in order of difficulty of handling. We will discuss the hyperspherical harmonics (HH) expansion method (HHEM) for an *ab initio* solution of the three-body Schrödinger equation. Details of this technique can be found in the lecture notes [1] and review articles [2, 3]. Even though this method can be formulated for an arbitrary but finite number of particles, we will discuss the simplest case of three particles in this chapter. In the next chapter we will see how this technique can be generalized for a larger number of interacting particles. In doing so, we will see the beauty of the mathematical process of going from an A-body system to the next $(A + 1)$-body system.

In HHEM the three-body relative wave function is expanded in the complete set of HH, which are the six-dimensional generalization of ordinary (three-dimensional) spherical harmonics. This method reduces the Schrödinger equation for the relative motion into a set of coupled differential equations in one variable, called the hyperradius. Analytic expressions for the HH are given in closed forms, in terms of standard orthogonal polynomials. Using these, a major part of coupling matrix element can be calculated analytically, leaving a simpler integral to be evaluated numerically. Hyperspherical formalism was originally introduced by Zernike and Brinkman in 1936 [4]

© Springer India 2016

T.K. Das, *Hyperspherical Harmonics Expansion Techniques*,
Theoretical and Mathematical Physics, DOI 10.1007/978-81-322-2361-0_3

and developed by Delves [5] and Smith [6] in the early 1960s. Ballot and Fabre [7] applied the method to the trinucleon system in the early 1980s. Since then this method has been developed to its full potential, including some modifications [8–12]. In this chapter, we will closely follow the approach of Ballot and Fabre [7].

An alternative basis, which incorporates the required symmetry of the three-body system was introduced by Simonov [13]. However, this procedure cannot be generalized to larger number of particles in an obvious manner. Moreover the convergence rate of this expansion was found to be too slow to be convenient for realistic potentials. The Ballot–Fabre approach, together with optimal subset and potential basis (see Sect. 4.5 of Chap. 4), is more convenient. As a result the Simonov basis has been used only rarely [14]. Hence we will not discuss the details of this method.

For simplicity, we consider three identical particles of mass m each, at position vectors $\vec{r}_1, \vec{r}_2, \vec{r}_3$, and interacting through mutual two-body forces. The treatment can easily be generalized for particles of different masses (hence nonidentical). There are many such systems in nature. Examples are: the trinucleon systems, ^{3}H and ^{3}He (neutron and proton are identical particles when isospin is taken into account), trimers of inert gases like He, Ne, Ar, etc. The Schrödinger equation for the system is

$$\left[-\sum_{i=1}^{3} \frac{\hbar^2}{2m} \nabla_i^2 + \sum_{j>i=1}^{3} V(\vec{r}_i - \vec{r}_j) \right] \psi(\vec{r}_1, \vec{r}_2, \vec{r}_3) = E_T \psi(\vec{r}_1, \vec{r}_2, \vec{r}_3). \tag{3.1}$$

When the interactions are mutual and there are no external forces, the CM motion can be separated. There are several possibilities for introducing the relative vectors. We will use the Jacobi coordinates, for which the Jacobian of the transformation is unity.

3.1 Jacobi Coordinates

To separate the center of mass (CM) and relative motions, we introduce the following system of vectors, called *Jacobi coordinates*

$$\vec{\xi}_1 = \vec{r}_2 - \vec{r}_1 = \vec{r}_{12}$$
$$\vec{\xi}_2 = \frac{2}{\sqrt{3}} [\vec{r}_3 - \frac{1}{2}(\vec{r}_1 + \vec{r}_2)] = \sqrt{3}(\vec{r}_3 - \vec{R})$$
$$\vec{R} = \frac{1}{3}(\vec{r}_1 + \vec{r}_2 + \vec{r}_3). \tag{3.2}$$

$\vec{R}$ is the CM coordinate and $\vec{\xi}_1, \vec{\xi}_2$ describe the relative motion. Note that the set of Jacobi vectors is not unique. Equation (3.2) is for the 'partition' $\{(12)3\}$. The partition $\{(ij)k\}$ (with i, j, k a cyclic permutation of $1, 2, 3$) is defined to be that set of Jacobi coordinates for which $\vec{\xi}_1 = \vec{r}_{ij} \equiv \vec{r}_j - \vec{r}_i$ and $\vec{\xi}_2 = \sqrt{3}(\vec{r}_k - \vec{R})$. Separation of the kinetic energy (KE) term into the KE of the relative motion and the KE of the CM motion follows from

$$\frac{1}{2}\sum_{i=1}^{3} \nabla_{\vec{r}_i}^2 = \sum_{j=1}^{2} \nabla_{\xi_j}^2 + \frac{1}{6}\nabla_{\vec{R}}^2. \tag{3.3}$$

Above equation can easily be verified using partial derivatives and Eq. (3.2). Note that the Jacobian of the transformation $\{\vec{r}_1, \vec{r}_2, \vec{r}_3\} \rightarrow \{\vec{\xi}_1, \vec{\xi}_2, \vec{R}\}$ is unity. Substitution of Eq. (3.3) in Eq. (3.1) separates the latter into a Schrödinger equation for the relative motion (depending on $\vec{\xi}_1, \vec{\xi}_2$ only) and a Schrödinger equation for the motion of the center of mass (depending on $\vec{R}$ only)

$$\left[-\frac{\hbar^2}{m}\left(\nabla_{\vec{\xi}_1}^2 + \nabla_{\vec{\xi}_2}^2 \right) + V(\vec{\xi}_1, \vec{\xi}_2) \right]\Psi(\vec{\xi}_1, \vec{\xi}_2) = E\Psi(\vec{\xi}_1, \vec{\xi}_2)$$

$$\left[-\frac{\hbar^2}{2M}\nabla_{\vec{R}}^2 \right]\Phi(\vec{R}) = E_{CM}\Phi(\vec{R}), \tag{3.4}$$

where E and E_{CM} are the energies of relative motion and CM motion respectively, $E_T = E + E_{CM}$ is the total energy and $\psi(\vec{r}_1, \vec{r}_2, \vec{r}_3) = \Psi(\vec{\xi}_1, \vec{\xi}_2)\Phi(\vec{R})$. Note that all the relative separations of the three particles can be expressed in terms of $\vec{\xi}_1$ and $\vec{\xi}_2$. The second of Eq. (3.4) shows that the CM moves as a free particle of the total mass $M = 3m$ and the first equation describes the relative motion of the three mutually interacting particles. This can be understood, as there is no external force on the system.

Solution of the CM equation is trivial. Thus the problem reduces to a solution of the Schrödinger equation for the relative motion. This is similar to the two-body system, where the relative motion is described by *one* three-dimensional relative vector $\vec{r}$. For the three-body system, we need *two* three-dimensional relative vectors $\vec{\xi}_1, \vec{\xi}_2$. These two vectors have a total of six components, which together define a *six-dimensional* vector space. In analogy with the three-dimensional spherical polar coordinates, we define a six-dimensional *hyperspherical polar coordinates* (which is simply called the 'hyperspherical coordinates') constituted by a hyperradius

$$\xi = \sqrt{\xi_1^2 + \xi_2^2} \tag{3.5}$$

and a set of five 'hyperangles', consisting of the four spherical polar angles: (ϑ_1, φ_1) of $\vec{\xi}_1$ and (ϑ_2, φ_2) of $\vec{\xi}_2$ and an angle ϕ defining the relative lengths of $\vec{\xi}_1, \vec{\xi}_2$ through

$$\xi_1 = \xi \sin\phi$$
$$\xi_2 = \xi \cos\phi, \tag{3.6}$$

with $0 \le \phi \le \frac{\pi}{2}$. The set of five hyperangles is denoted by $\Omega_6 \equiv \{\vartheta_1, \varphi_1, \vartheta_2, \varphi_2, \phi\}$. They define the hyperangles in the six-dimensional (6D) space. The subscript of Ω denotes the dimension of the space spanned by the relative vectors.

Next, in analogy with the expression of the three-dimensional Laplace operator in spherical polar coordinates, Eq. (2.2), we express the six-dimensional Laplace operator in hyperspherical coordinates

$$\sum_{j=1}^{2} \nabla_{\xi_j}^2 = \left(\frac{\partial^2}{\partial \xi^2} + \frac{5}{\xi} \frac{\partial}{\partial \xi} - \frac{\mathcal{L}^2(\Omega_6)}{\hbar^2 \xi^2} \right), \tag{3.7}$$

where

$$\mathcal{L}^2(\Omega_6) = -\hbar^2 \left[\frac{\partial^2}{\partial \phi^2} + 4 \cot 2\phi \frac{\partial}{\partial \phi} - \frac{l_1^2(\hat{\xi}_1)}{\hbar^2 \sin^2 \phi} - \frac{l_2^2(\hat{\xi}_2)}{\hbar^2 \cos^2 \phi} \right] \tag{3.8}$$

is the six-dimensional 'hyperangular momentum' operator. Here $\hat{\xi}_i \equiv \{\vartheta_i, \varphi_i\}$ and $\vec{l}_i$ is the three-dimensional orbital angular momentum operator corresponding to $\hat{\xi}_i$, whose square is given by Eq. (2.4) ($i = 1, 2$). Note that all the angular momenta in Eqs. (3.7) and (3.8) are expressed in units of $\hbar$, as is commonly done. Thus eigenvalues of $\mathcal{L}^2$, l_1^2 and l_2^2 have a factor $\hbar^2$ (compare with Eqs. (2.2)–(2.4)). For simplicity of expressions, from now on, we choose 'theoretical units,' in which $\hbar = 1$. Also note that in Ref. [7], all square of angular momentum operators are taken with negative signs. We follow in Eqs. (2.2)–(2.4) and (3.7)–(3.8) the usual sign convention with positive eigenvalues. The final results will be the same, if one is consistent in the signs.

3.2 Hyperspherical Harmonics

In analogy with the case of a single relative vector, where we expressed the wave function in terms of the spherical harmonics, we expand the relative wave function $\Psi(\vec{\xi}_1, \vec{\xi}_2)$ of Eq. (3.4) in the 'hyperspherical harmonics.' The latter are defined as the eigenfunctions $\mathcal{Y}(\Omega_6)$ of square of the hyperangular momentum operator (as spherical harmonics are the eigenfunctions of square of the three-dimensional (3D) orbital angular momentum operator)

$$\mathcal{L}^2(\Omega_6)\mathcal{Y}(\Omega_6) = \lambda \mathcal{Y}(\Omega_6), \tag{3.9}$$

where λ is the eigen value. Note that the solution of the 3D Laplace equation using the 3D Laplace operator, Eq. (2.3), is

$$\nabla^2 \{r^l Y_{lm}(\hat{r})\} = 0,$$

where Eq. (2.5) has been used. Now $r^l Y_{lm}(\hat{r})$ is a homogeneous polynomial of degree l in the Cartesian components of $\vec{r}$. In a similar fashion, the solution of the 6D Laplace equation involving the 6D Laplace operator (Eq. (3.7)), is a homogeneous polynomial of (say) degree L in the Cartesian components of $(\vec{\xi}_1, \vec{\xi}_2)$. From the definition of hyperangles including Eq. (3.6), it is clear that this homogeneous polynomial has the form of ξ^L times a function of hyperangles $(\hat{\xi}_1, \hat{\xi}_2, \phi)$, denoted by $\mathcal{Y}_{[L]}(\Omega_6)$ and called 'hyperspherical harmonics' of order L. Using Eq. (3.7) in the 6D Laplace equation

$$\sum_{j=1}^{2} \nabla^2_{\xi_j} \{\xi^L \mathcal{Y}_{[L]}(\Omega_6)\} = 0, \tag{3.10}$$

we see that Eq. (3.9) is satisfied with $\lambda = L(L+4)$

$$\mathcal{L}^2(\Omega_6)\mathcal{Y}_{[L]}(\Omega_6) = L(L+4)\mathcal{Y}_{[L]}(\Omega_6). \tag{3.11}$$

Here $[L]$ denotes the set of necessary quantum numbers for a given value of L. Now Ω_6 is a set of five variables. Hence we need five quantum numbers. Next from Eq. (3.8), it is clear that the variables (ϑ_1, φ_1) and (ϑ_2, φ_2) separate. Hence their eigenfunctions are the standard spherical harmonics $Y_{l_1 m_1}(\vartheta_1, \varphi_1)$ and $Y_{l_2 m_2}(\vartheta_2, \varphi_2)$ respectively, with quantum numbers (l_1, m_1) and (l_2, m_2). Hence

$$\mathcal{Y}_{[L]}(\Omega_6) = {}^{(2)}\mathcal{P}_L^{l_2, l_1}(\phi)\, Y_{l_1 m_1}(\vartheta_1, \varphi_1) Y_{l_2 m_2}(\vartheta_2, \varphi_2), \tag{3.12}$$

where ${}^{(2)}\mathcal{P}_L^{l_2, l_1}(\phi)$ is still to be determined. The left-superscript (2) indicates the number of relative vectors. It is introduced in view of the generalization to A-body problem, to be discussed in Chap. 4. The symbol $[L]$ represents the five quantum numbers $\{l_1, m_1, l_2, m_2, L\}$ in the uncoupled basis. If the interaction potential is spherically symmetric, the total angular momentum $\vec{l} = \vec{l}_1 + \vec{l}_2$ is a good quantum number of the system and in the coupled basis the symbol $[L]$ represents $\{l_1, l_2, l, m_l, L\}$, where m_l is the projection of $\vec{l}$. Substitution of Eq. (3.12) in Eq. (3.11), together with Eq. (3.8) gives (note that we are now using the units in which $\hbar = 1$)

$$\left[-\frac{\partial^2}{\partial \phi^2} - 4\cot 2\phi \frac{\partial}{\partial \phi} + \frac{l_1(l_1+1)}{\sin^2 \phi} + \frac{l_2(l_2+1)}{\cos^2 \phi} \right] {}^{(2)}\mathcal{P}_L^{l_2, l_1}(\phi) = L(L+4)\, {}^{(2)}\mathcal{P}_L^{l_2, l_1}(\phi). \tag{3.13}$$

This is the eigenvalue equation satisfied by ${}^{(2)}\mathcal{P}_L^{l_2, l_1}(\phi)$. This can be put in a standard form by writing

$${}^{(2)}\mathcal{P}_L^{l_2, l_1}(\phi) = (\cos \phi)^{l_2} (\sin \phi)^{l_1} g(\phi), \tag{3.14}$$

and changing the variable ϕ to a new variable $x = \cos 2\phi$. With these changes, Eq. (3.13) becomes the standard Jacobi differential equation [15]

$$(1-x^2)g''(x) + [\beta - \alpha - (\alpha + \beta + 2)x]g'(x) + n(n + \alpha + \beta + 1)g(x) = 0, \tag{3.15}$$

where $\alpha = l_1 + \frac{1}{2}, \beta = l_2 + \frac{1}{2}$ and n is associated with the eigenvalue $n(n + \alpha + \beta + 1)$ through

$$L = 2n + l_1 + l_2. \tag{3.16}$$

We notice that Eq. (3.15) in the interval $[-1, 1]$ can be put in the Sturm–Liouville form [16] by multiplying it through by $(1-x)^\alpha (1+x)^\beta$. Hence we can identify the eigenvalue as $n(n + \alpha + \beta + 1)$ and the weight function as

$$w(x) = (1-x)^\alpha (1+x)^\beta. \tag{3.17}$$

The solution of Eq. (3.15) is the standard Jacobi function. The requirement of regularity of $g(x)$ at $x = \pm 1$ is satisfied when n is a nonnegative integer $n = 0, 1, 2, \ldots$ Then the solution of Eq. (3.15) becomes the *Jacobi polynomials* [15]

$$g(x) = P_n^{\alpha,\beta}(x). \tag{3.18}$$

The quantity L of Eq. (3.16) is called the 'grand orbital quantum number.'

The Sturm–Liouville theory [16] guarantees that the eigenfunctions of Eq. (3.15) belonging to different eigenvalues are orthogonal with respect to the weight function in the interval $[-1, 1]$

$$\int_{-1}^{1} P_n^{\alpha,\beta}(x) P_{n'}^{\alpha,\beta}(x)(1-x)^\alpha (1+x)^\beta dx = h_n^{\alpha,\beta} \delta_{n,n'}, \tag{3.19}$$

where $h_n^{\alpha,\beta}$ is the 'norm' of standard Jacobi polynomials [15] given by

$$h_n^{\alpha,\beta} = \frac{2^{\alpha+\beta+1}}{2n+\alpha+\beta+1} \frac{\Gamma(n+\alpha+1)\Gamma(n+\beta+1)}{\Gamma(n+1)\Gamma(n+\alpha+\beta+1)}. \tag{3.20}$$

Here $\Gamma(x) = (x+1)!$ is the standard gamma function [16].

The volume element in the 6D space spanned by $(\vec{\xi}_1, \vec{\xi}_2)$ is

$$
\begin{aligned}
dV_6 &= d^3\xi_1 d^3\xi_2 \\
&= \xi_1^2 d\xi_1 \sin\vartheta_1 d\vartheta_1 d\varphi_1 \xi_2^2 d\xi_2 \sin\vartheta_2 d\vartheta_2 d\varphi_2
\end{aligned} \tag{3.21}
$$

Using Eq. (3.6), the 6D volume element in terms of the hyperspherical variables becomes

$$dV_6 = \xi^5 d\xi \cos^2\phi \sin^2\phi d\phi \sin\vartheta_1 d\vartheta_1 d\varphi_1 \sin\vartheta_2 d\vartheta_2 d\varphi_2. \tag{3.22}$$

The intervals for ξ and ϕ are $[0, \infty]$ and $[0, \pi/2]$ respectively. The intervals for ϑ_i and φ_i are the usual ones, $viz.$ $[0, \pi]$ and $[0, 2\pi]$ respectively for $i = 1, 2$. Writing $dV_6 = \xi^5 d\xi d\Omega_6$, we have

$$d\Omega_6 = \cos^2 \phi \, \sin^2 \phi \, d\phi \, \sin \vartheta_1 d\vartheta_1 d\varphi_1 \, \sin \vartheta_2 d\vartheta_2 d\varphi_2. \qquad (3.23)$$

The HH are normalized according to

$$\int \mathcal{Y}^*_{[L]}(\Omega_6) \mathcal{Y}_{[L']}(\Omega_6) d\Omega_6 = \delta_{[L],[L']}$$

$$= \delta_{n,n'} \delta_{l_1,l_1'} \delta_{m_1,m_1'} \delta_{l_2,l_2'} \delta_{m_2,m_2'}. \qquad (3.24)$$

The ${}^{(2)}\mathcal{P}^{l_2,l_1}_L(\phi)$ functions are orthonormalized according to

$$\int_0^{\frac{\pi}{2}} {}^{(2)}\mathcal{P}^{l_2,l_1}_L(\phi) \, {}^{(2)}\mathcal{P}^{l_2,l_1}_{L'}(\phi) \, \sin^2 \phi \, \cos^2 \phi \, d\phi = \delta_{L,L'}. \qquad (3.25)$$

Using Eqs. (3.14), (3.16), (3.18)–(3.20) in Eq. (3.25), we have for the complete expression for normalized ${}^{(2)}\mathcal{P}^{l_2,l_1}_L(\phi)$

$${}^{(2)}\mathcal{P}^{l_2,l_1}_L(\phi) = N^{l_2,l_1}_L (\cos \phi)^{l_2} (\sin \phi)^{l_1} P_n^{l_1+\frac{1}{2},l_2+\frac{1}{2}}(\cos 2\phi),$$

where

$$N^{l_2,l_1}_L = \left[\frac{2(L+2)\Gamma(L+2-n)\Gamma(n+1)}{\Gamma(n+l_1+\frac{3}{2})\Gamma(n+l_2+\frac{3}{2})} \right]^{\frac{1}{2}}. \qquad (3.26)$$

The complete expression of 6D HH (for three-body systems) is given by Eqs. (3.12) and (3.26). The Sturm–Liouville theory guarantees that the set of all such functions forms a complete set in 6D angular hyperspace, just as the set of all ordinary spherical harmonics forms a complete set in 3D polar angular space. Thus the HH basis forms a complete basis for expansion of any function in the 6D space, in particular for the relative wave function of the three-body system. In the above, we have obtained these basis functions, without any consideration for the symmetry. However, symmetry of the wave function is an important aspect for systems containing two or more identical particles. We will discuss construction of the basis with different symmetries in Chap. 4.

The HH are used in the theoretical ab $initio$ studies of few-body systems ranging from fundamental particles, nuclei, atoms up to molecules and clusters. This method is quite popular in physics research due to its ab $initio$ approach. Even systems which are not strictly few-body can be modeled as few-body with one or more cluster of particles treated as a single particle, as we will see in Sect. 4.6 of Chap. 4. Although the hyperspherical technique is a highly mathematical one, it has recently been used in medical research as well [17].

3.3 Schrödinger Equation for Relative Motion

The first of Eq. (3.4) describes the relative motion of three equal mass particles, ignoring spin and isospin degrees of freedom, i.e., in space variables only. As we discussed earlier, we expand the relative wave function in the complete set of 6D HH. The expansion coefficients become functions of the hyperradius ξ. In order to remove first derivative with respect to ξ in the hyperradial Schrödinger equation, we introduce a factor of $\xi^{-\frac{5}{2}}$ (note that in the 3D case, this factor was r^{-1})

$$\Psi(\vec{\xi}_1, \vec{\xi}_2) = \sum_{[L]} \xi^{-\frac{5}{2}} u_{[L]}(\xi) \mathcal{Y}_{[L]}(\Omega_6). \tag{3.27}$$

Substituting this in the relative Schrödinger equation (first of Eq. (3.4)), using Eq. (3.7), we get

$$\sum_{[L']} \left[-\frac{\hbar^2}{m} \left(\frac{\partial^2}{\partial \xi^2} + \frac{5}{\xi} \frac{\partial}{\partial \xi} - \frac{\mathcal{L}^2(\Omega_6)}{\xi^2} \right) + V(\xi, \Omega_6) - E \right] \xi^{-\frac{5}{2}} u_{[L']}(\xi) \mathcal{Y}_{[L']}(\Omega_6) = 0. \tag{3.28}$$

Substituting Eq. (3.11), we have

$$\sum_{[L']} \left[-\frac{\hbar^2}{m} \left(\frac{d^2}{d\xi^2} - \frac{L'(L'+4)}{\xi^2} \right) - E \right] u_{[L']}(\xi) \mathcal{Y}_{[L']}(\Omega_6)$$

$$+ \sum_{[L']} V(\xi, \Omega_6) u_{[L']}(\xi) \mathcal{Y}_{[L']}(\Omega_6) = 0. \tag{3.29}$$

We project this equation on a particular HH by premultiplying with $\mathcal{Y}^*_{[L]}(\Omega_6)$, integrating over Ω_6 and using the orthonormality of HH, Eq. (3.24)

$$\left[-\frac{\hbar^2}{m} \left(\frac{d^2}{d\xi^2} - \frac{L(L+4)}{\xi^2} \right) - E \right] u_{[L]}(\xi) + \sum_{[L']} V_{[L],[L']}(\xi) u_{[L']}(\xi) = 0, \tag{3.30}$$

where $V_{[L],[L']}(\xi)$ is the potential matrix element (PME). It is a function of ξ and is given by

$$V_{[L],[L']}(\xi) = \int \mathcal{Y}^*_{[L]}(\Omega_6) V(\xi, \Omega_6) \mathcal{Y}_{[L']}(\Omega_6) d\Omega_6. \tag{3.31}$$

Equation (3.30) is a set of coupled differential equations (CDE) in the hyperradial variable ξ, where $u_{[L]}(\xi)$ are called the *hyperspherical partial waves*. In order to solve a given three-body problem, one has to calculate the potential matrix elements using Eq. (3.31) and use them in the CDE, Eq. (3.30), subject to appropriate boundary conditions: for a bound state, $u_{[L]}(\xi)$ must vanish for both $\xi \to 0$ and $\xi \to \infty$. For a given potential with a known asymptotic behavior in these limits, the asymptotic functional forms of $u_{[L]}(\xi)$ can be given in analytic forms, for a better accuracy and faster convergence. The CDE can in general be solved numerically exactly by the renormalized Numerov algorithm [18]. However, this procedure requires a large amount of computation. Alternately, one can solve the set of CDE by the hyperspherical adiabatic approximation, which gives a reasonable solution for most problems in physics, at a much reduced computation level. We will discuss the hyperspherical adiabatic approximation in Chap. 10.

3.4 Calculation of Potential Matrix Element

The main difficulty in the solution of the three-body problem is the computation of the PME. Eq. (3.31) is in general a five-dimensional integral and the integrand contains $V(\xi, \Omega_6) = V(\vec{r}_{12}) + V(\vec{r}_{23}) + V(\vec{r}_{31})$, where $\vec{r}_{ij} = \vec{r}_j - \vec{r}_i$. From Eq. (3.2), we have

$$\vec{r}_{12} = \vec{\xi}_1$$
$$\vec{r}_{32} = \frac{1}{2}\vec{\xi}_1 - \frac{\sqrt{3}}{2}\vec{\xi}_2$$
$$\vec{r}_{13} = \frac{1}{2}\vec{\xi}_1 + \frac{\sqrt{3}}{2}\vec{\xi}_2. \tag{3.32}$$

Thus even if the two-body potentials are central, $V(\xi, \Omega_6)$ depends on the angle between $\vec{\xi}_1$ and $\vec{\xi}_2$. In this case, the three-body system is invariant under overall rotations and the total orbital angular momentum $\vec{l}$ (where $\vec{l} = \vec{l}_1 + \vec{l}_2$) and its projection m_l are good quantum numbers. Hence we take the expansion basis as the coupled HH basis. An angular momentum coupled HH is constructed from Eq. (3.12)

$$\mathcal{Y}_{l_1,l_2,l,m_l,L}(\Omega_6) = {}^{(2)}\mathcal{P}_L^{l_2,l_1}(\phi)\left[Y_{l_1 m_1}(\vartheta_1, \varphi_1)Y_{l_2 m_2}(\vartheta_2, \varphi_2)\right]_{l m_l}, \tag{3.33}$$

where $\left[\cdots\right]_{l m_l}$ indicates angular momentum coupling, *viz.*

$$\left[Y_{l_1 m_1}(\vartheta_1, \varphi_1)Y_{l_2 m_2}(\vartheta_2, \varphi_2)\right]_{l m_l} = \sum_{m_1,m_2} \langle l_1, m_1, l_2, m_2 | l, m_l \rangle$$
$$\times Y_{l_1 m_1}(\vartheta_1, \varphi_1)Y_{l_2 m_2}(\vartheta_2, \varphi_2). \tag{3.34}$$

Here $\langle l_1, m_1, l_2, m_2 | l, m_l \rangle$ is a Clebsch–Gordan coefficient. Thus, although the calculation of PME for the pair (12) is simple (which reduces to a one-dimensional integral over the variable ϕ only), those for the pairs (23) and (31) are very complicated. In the next subsection, we will discuss a relatively simple way to evaluate the PME.

3.4.1 Expansion of potential in hyperspherical multipoles

The total interaction potential $V(\xi, \Omega_6)$ is, in general, a function of the hyperangles, besides the hyperradius. Hence for a fixed value of ξ it can be expanded in the complete set of HH

$$V(\xi, \Omega_6) = \sum_{[L'']} v_{[L'']}(\xi) \mathcal{Y}_{[L'']}(\Omega_6). \tag{3.35}$$

The expansion coefficient, $v_{[L'']}(\xi)$, is a function of ξ and is called the hyperspherical potential multipole (PM). Substituting Eq. (3.35) in Eq. (3.31)

$$V_{[L],[L']}(\xi) = \sum_{[L'']} v_{[L'']}(\xi) \langle [L] | [L''] | [L'] \rangle, \tag{3.36}$$

where

$$\langle [L] | [L''] | [L'] \rangle = \int \mathcal{Y}^*_{[L]}(\Omega_6) \mathcal{Y}_{[L'']}(\Omega_6) \mathcal{Y}_{[L']}(\Omega_6) d\Omega_6 \tag{3.37}$$

is called the geometrical structure coefficient (GSC). Note that although the GSC is a five-dimensional integral, it is independent of the potential and ξ. The full set of GSC needs to be calculated only once and stored. These stored GSC can be used in Eq. 3.36 to calculate the potential matrix element for all values of ξ, just by calculating the potential multipole for each value of ξ. Calculation of the matrix element of different potentials can also be done using the same set of GSCs. This reduces the bulk of the calculations. Now this will be useful, if the sum over $[L'']$ in Eq. (3.36) is a finite one. Indeed this is the case, as can be seen from the following. One notices that $\xi^L \mathcal{Y}_{[L]}(\Omega_6)$ is a homogeneous harmonic polynomial of degree L in the Cartesian components of $\vec{\xi}_1$ and $\vec{\xi}_2$. Multiply Eq. (3.37) by $\xi^{L+L'}$ on both sides and note that $\xi^L \mathcal{Y}_{[L]} \times \xi^{L'} \mathcal{Y}_{[L']}$ on the right side is a polynomial of degrees $L + L'$. Hence this product can be expanded in terms of $\xi^{L'''} \mathcal{Y}_{[L''']}$, with values of L''' running from 0 to $L + L'$. Then, using the orthogonality of $\mathcal{Y}_{[L'']}$ in the integral on the right side of Eq. (3.37), we see that $\langle [L] | [L''] | [L'] \rangle$ vanishes unless $L'' \le (L + L')$. Combining each pair of HH within the integrand of Eq. (3.37), we finally have

$$L'' \le (L + L')$$
$$L' \le (L + L'')$$
$$L \le (L' + L'').$$

These define the *triangle rule*, in which L, L' and L'' form the three sides of a triangle. This selection rule can also be written as

$$\langle [L] | [L''] | [L'] \rangle = 0$$
$$\text{unless} \quad |L - L'| \le L'' \le (L + L') \tag{3.38}$$

This is the same as the angular momentum selection rule. Note that L, L' and L'' are all nonnegative integers. Thus only a finite number of L'' values [from $|L - L'|$ to $(L + L')$] are needed in Eq. (3.36). Potential multipoles corresponding to only these values of L'' are to be evaluated for given L, L' and ξ. In the next subsection, we will see how these can be evaluated for a central potential. Still the calculation of each GSC remains a difficult task, since it involves a five-dimensional integral. We will discuss in Sect. 3.6 how the GSC can be evaluated by solving a set of linear inhomogeneous equations, *without even doing a single integration*.

3.4.2 Calculation of potential multipole

The next problem is the evaluation of the potential multipole. Multiplying Eq. (3.35) by $\mathcal{Y}^*_{[L]}(\Omega_6)$, integrating over Ω_6 and using the orthonormality of HH, we have

$$v_{[L]}(\xi) = \int V(\xi, \Omega_6) \mathcal{Y}^*_{[L]}(\Omega_6) d\Omega_6. \tag{3.39}$$

Here $V(\xi, \Omega_6)$ is the sum of three pairwise interactions. For central two-body interactions

$$V(\xi, \Omega_6) = V^{(12)}(r_{12}) + V^{(23)}(r_{23}) + V^{(31)}(r_{31}), \tag{3.40}$$

where $V^{(ij)}(r_{ij})$ is the interaction between particles labeled i and j. If the particles are identical, all $V^{(ij)}$ are the same (say V). Using Eq. (3.32) we see that calculation of potential multipole of the (12)-pair is easy

$$
\begin{aligned}
v^{(12)}_{[L]}(\xi) &= \int V(\xi_1) \mathcal{Y}^*_{[L]}(\Omega_6) d\Omega_6. \\
&= \delta_{l_1,0}\delta_{l_2,0}\delta_{l,0}\delta_{m_l,0} \, 4\pi \int_0^{\pi/2} V(\xi \cos\phi) \, {}^{(2)}P^{0,0}_{2K}(\phi) \cos^2\phi \sin^2\phi \, d\phi \\
&= \delta_{l_1,0}\delta_{l_2,0}\delta_{l,0}\delta_{m_l,0} \, 8\pi \frac{\Gamma(K+2)}{\Gamma(K+\frac{3}{2})} \int_0^{\pi/2} V(\xi \cos\phi) P_K^{\frac{1}{2},\frac{1}{2}}(\cos 2\phi) \cos^2\phi \sin^2\phi \, d\phi \\
&= \delta_{l_1,0}\delta_{l_2,0}\delta_{l,0}\delta_{m_l,0} \, 8\pi \frac{\Gamma(K+2)}{\Gamma(K+\frac{3}{2})} \int_{-1}^{1} V\left(\xi\sqrt{\frac{1+z}{2}}\right) P_K^{\frac{1}{2},\frac{1}{2}}(z) \sqrt{1-z^2} \, dz \quad (3.41)
\end{aligned}
$$

where, in going from the first to the second line we used Eqs. (3.6), (3.33), (3.16) (replacing n by K) and the orthonormality of spherical harmonics and in going from the second to the third line we used Eq. (3.26). To get the last line, $z = \cos 2\phi$ has been substituted. This is a one-dimensional integral and easy to calculate by a suitable numerical quadrature.

Calculation of the potential multipoles for the (23) and the (31) pairs are not so easy, since both of them will involve ξ_1, ξ_2 and the angle between $\vec{\xi}_1$ and $\vec{\xi}_2$. However, for three spin zero bosons, the total wave function should be symmetric under any pair exchange. Hence $\mathcal{Y}_{[L]}(\Omega_6)$ should be replaced by $\mathcal{Y}_{[L]}^{(S)}(\Omega_6)$, which is symmetric under exchange of any pair. We will discuss in Sect. 3.5 how to construct this symmetric HH basis for three identical particles. In this symmetric basis, the contribution of each of the three pairs is the same. Hence

$$v_{[L]}(\xi) = 3v_{[L]}^{(12)}(\xi),\tag{3.42}$$

where we calculate $v_{[L]}^{(12)}(\xi)$ replacing $\mathcal{Y}_{[L]}^*(\Omega_6)$ by a symmetric HH $\mathcal{Y}_{[L]}^{(S)*}(\Omega_6)$ in the first line of Eq. (3.41).

3.5 Symmetrization of HH Basis

It is easy to see that the HH is symmetric (or antisymmetric) under $1 \longleftrightarrow 2$, if l_1 is restricted to even (or odd) values. To impose the symmetries with respect to exchange of other pairs, we introduce a 3D vector $\vec{z}(\varpi)$ as a linear combination of $\vec{\xi}_1$ and $\vec{\xi}_2$, depending on the parameter ϖ

$$\vec{z}(\varpi) = \sin\varpi\,\vec{\xi}_1 + \cos\varpi\,\vec{\xi}_2.\tag{3.43}$$

Then comparing with Eqs. (3.2) and (3.32), we have

$$\vec{r}_2 - \vec{r}_1 = \vec{z}(\pi/2)$$
$$\vec{r}_3 - \vec{r}_2 = \vec{z}(\pi/2 - 2\pi/3)$$
$$\vec{r}_1 - \vec{r}_3 = \vec{z}(\pi/2 + 2\pi/3),\tag{3.44}$$

as also the position vectors of the particles with respect to the CM are given by

$$\vec{r}_3 - \vec{R} = \frac{1}{\sqrt{3}}\vec{z}(0)$$

$$\vec{r}_1 - \vec{R} = \frac{1}{\sqrt{3}}\vec{z}(-2\pi/3)$$

$$\vec{r}_2 - \vec{R} = \frac{1}{\sqrt{3}}\vec{z}(2\pi/3).\tag{3.45}$$

Thus, from Eq. (3.2) we see that for $\varpi = \pi/2$, $\pi/2 - 2\pi/3$ and $\pi/2 + 2\pi/3$ the pair of 3D vectors $\{\vec{z}(\varpi), \vec{z}(\varpi - \pi/2)\}$ are the $\{\vec{\xi}_1, \vec{\xi}_2\}$ vectors for partitions $\{(12)3\}$, $\{(23)1\}$ and $\{(31)2\}$ respectively. For this reason $\vec{z}(\varpi)$ is called the 'kinematic rotation vector.' The relative wavefunction Ψ of Eq. (3.4) is $\Psi\left(\vec{z}(\varpi), \vec{z}(\varpi - \pi/2)\right)$ for $\varpi = \pi/2$, $\pi/2 - 2\pi/3$ and $\pi/2 + 2\pi/3$, respectively corresponding to the three partitions $\{(12)3\}$, $\{(23)1\}$ and $\{(31)2\}$. Define a symmetrizing operator Σ_0 by

$$\Sigma_0 f(\varpi) = \frac{1}{3} \sum_{\varpi = \pi/2, \pi/2 - 2\pi/3, \pi/2 + 2\pi/3} f(\varpi). \tag{3.46}$$

Then

$$\Psi^{(S)} = \Sigma_0 \Psi \left(\vec{z}(\varpi), \vec{z}(\varpi - \pi/2) \right) \tag{3.47}$$

is the totally symmetric combination under any pair exchange, when l_1 is restricted to even nonnegative integers. Note that the hyperradius ξ is invariant under the permutations of the three particles. We expand $\Psi \left(\vec{z}(\varpi), \vec{z}(\varpi - \pi/2) \right)$ as in Eq. (3.27)

$$\Psi \left(\vec{z}(\varpi), \vec{z}(\varpi - \pi/2) \right) = \sum_{[L]} \xi^{-5/2} u_{[L]}(\xi) \mathcal{Y}_{(l_1, l_2) l m_l L}(\Omega_{6, \varpi}). \tag{3.48}$$

Now, $\mathcal{Y}_{(l_1, l_2) l m_l L}(\Omega_{6, \varpi})$ is a function in the 6D hyperangular space and can be expanded in the set of $\mathcal{Y}(\Omega_6)$ corresponding to the $\{(12)3\}$ partition

$$\mathcal{Y}_{(l_1, l_2) l m_l L}(\Omega_{6, \varpi}) = \mathcal{N} \sum_{l_1', l_2'} \mathcal{A}_{l_1', l_2'}^{[L]}(\varpi) \mathcal{Y}_{(l_1', l_2') l m_l L}(\Omega_6). \tag{3.49}$$

Note that the grand orbital (hyperangular momentum) quantum number L, associated with the hyperradius does not change and the sum is over $l_1' l_2'$ only. The coefficients $\mathcal{A}_{l_1', l_2'}^{[L]}(\varpi)$ are called Raynal–Revai coefficients [19] and $\mathcal{N}$ is a normalization constant. The Raynal–Revai coefficients are given by [7]

$$\mathcal{A}_{l_1', l_2'}^{[L]}(\varpi) = \frac{\pi}{2} \sum_{\Lambda \Lambda' \lambda_1' \lambda_2' \lambda_1 \lambda_2} (-1)^{n' + \lambda_2'} \frac{(L+4)!}{(L - \Lambda + 2)!(\Lambda + 2)!} (\cos \varpi)^{\Lambda'} (\sin \varpi)^{L - \Lambda'}$$

$$\cdot \begin{pmatrix} l_1 & \lambda_1 & \lambda_1' \\ l_2 & \lambda_2' & \lambda_2 \\ l & l_1' & l_2' \end{pmatrix} \cdot \Delta(\lambda_1 \lambda_1' l_1) \Delta(\lambda_2 \lambda_2' l_2)$$

$$\cdot \Delta(l_1' \lambda_1 \lambda_2') \Delta(l_2' \lambda_2 \lambda_1') \langle {}^{(2)}\mathcal{P}_\Lambda^{\lambda_2 \lambda_1} | {}^{(2)}\mathcal{P}_{\Lambda'}^{\lambda_2' \lambda_1'} | {}^{(2)}\mathcal{P}_L^{l_2 l_1} \rangle \langle {}^{(2)}\mathcal{P}_L^{l_2' l_1'} | {}^{(2)}\mathcal{P}_{L'}^{l_1 l_2} | {}^{(2)}\mathcal{P}_{\Lambda'}^{\lambda_1' \lambda_2'} \rangle. \tag{3.50}$$

Here the 3×3 matrix symbol is a 9-j symbol and

$$\Delta(abc) = [(2a+1)(2b+1)(2c+1)]^{\frac{1}{2}} \begin{pmatrix} a & b & c \\ 0 & 0 & 0 \end{pmatrix}. \tag{3.51}$$

The last symbol in Eq. (3.51) is a 3-j symbol. The 3-P coefficient $\langle {}^{(2)}\mathcal{P}_\Lambda^{\lambda_2 \lambda_1} | {}^{(2)}\mathcal{P}_{\Lambda'}^{\lambda_2' \lambda_1'} | {}^{(2)}\mathcal{P}_L^{l_2 l_1} \rangle$ in Eq. (3.50) is given by

$$\langle {}^{(2)}\mathcal{P}_\Lambda^{\lambda_2 \lambda_1} | {}^{(2)}\mathcal{P}_{\Lambda'}^{\lambda_2' \lambda_1'} | {}^{(2)}\mathcal{P}_L^{l_2 l_1} \rangle = \int_0^{\frac{\pi}{2}} d\phi \; {}^{(2)}\mathcal{P}_\Lambda^{\lambda_1 \lambda_2}(\phi) \; {}^{(2)}\mathcal{P}_{\Lambda'}^{\lambda_1' \lambda_2}(\phi) \; {}^{(2)}\mathcal{P}_L^{l_2 l_2}(\phi)$$
$$\times (\sin \phi)^2 (\cos \phi)^2. \tag{3.52}$$

Using Eq. (3.50) in Eq. (3.49), one can calculate the HH for partitions $\{(23)1\}$ and $\{(31)2\}$ and their substitutions in Eq. (3.39) give the corresponding potential multipoles. On the other hand, substitution of HH for the three partitions in Eq. (3.47), together with Eqs. (3.48)–(3.50) give the symmetric HH for the three-body system. Numerical calculation of the Raynal–Revai coefficients was performed by Khan et al. [20].

3.6 Calculation of GSC for Central Potentials

From Eqs. (3.36) and (3.41), we see that for central two-body potentials, the set of quantum numbers represented by $[L'']$ is restricted to $\{l_1'' = 0, l_2'' = 0, l'' = 0, m_l'' = 0, L'' = 2K''\}$, with K'' a nonnegative integer. This is true for the (ij)-pair interaction in the $(ij)k$ partition, with $i, j, k = 1, 2, 3$ cyclic permutation. For other pair interaction, the Raynal–Revai coefficients, Eq. (3.50) can be used. Hence the integrals over $d\hat{\xi}_1 d\hat{\xi}_2$ in Eq. (3.37) are simply done by orthonormality of spherical harmonics and we have

$$\langle [L]|[L'']|[L'] \rangle = \delta_{l_1,l_1'} \delta_{l_2,l_2'} \delta_{l,l'} \delta_{m_l,m_l'} \left({}^{(2)}\mathcal{P}_L^{l_2,l_1} \Big| {}^{(2)}\mathcal{P}_{L''}^{0,0} \Big| {}^{(2)}\mathcal{P}_{L'}^{l_2,l_1} \right). \tag{3.53}$$

Thus for central two-body potentials, the GSC reduces to the 3-P coefficients. Restricting ourselves to the $l = 0$ states of the three-body system (note that the ground state will be a $l = 0$ state), we have $l_1 = l_2$ and hence

$$\langle [L]|[L'']|[L'] \rangle = \left({}^{(2)}\mathcal{P}_{2K}^{l_1,l_1} \Big| {}^{(2)}\mathcal{P}_{2K''}^{0,0} \Big| {}^{(2)}\mathcal{P}_{2K'}^{l_1,l_1} \right), \tag{3.54}$$

where $[L] = \{l_1, l_1, 0, 0, 2K\}$, $[L'']=\{0, 0, 0, 0, 2K''\}$ and $[L'] = \{l_1, l_1, 0, 0, 2K'\}$, with nonnegative integers for K, K', K''. Substituting for the ${}^{(2)}\mathcal{P}$ functions from Eq. (3.26) in Eq. (3.52), this takes the form (writing l for l_1 for convenience)

$$\left({}^{(2)}\mathcal{P}_{2K}^{l,l} \Big| {}^{(2)}\mathcal{P}_{2K''}^{0,0} \Big| {}^{(2)}\mathcal{P}_{2K'}^{l,l} \right) = \int_0^{\frac{\pi}{2}} {}^{(2)}\mathcal{P}_{2K}^{l,l}(\phi) {}^{(2)}\mathcal{P}_{2K''}^{0,0}(\phi) {}^{(2)}\mathcal{P}_{2K'}^{l,l}(\phi) \sin^2\phi \cos^2\phi \, d\phi$$

$$= N_{2K}^{l,l} N_{2K''}^{0,0} N_{2K'}^{l,l} 2^{-(2l+3)} \int_{-1}^1 (1-x)^{l+\frac{1}{2}} (1+x)^{l+\frac{1}{2}}$$

$$\times P_{K-l}^{l+\frac{1}{2},l+\frac{1}{2}}(x) P_{K''}^{\frac{1}{2}\frac{1}{2}}(x) P_{K'-l}^{l+\frac{1}{2},l+\frac{1}{2}}(x) dx \tag{3.55}$$

We see that this is an integral of three Jacobi polynomials times the weight function of the Jacobi polynomials over the interval $[-1, 1]$. Now the second order differential operator corresponding to the Jacobi differential equation (3.15), when putting in the Sturm–Liouville form, is a Hermitian differential operator. Hence its eigenfunctions, the Jacobi polynomials, are orthogonal (see Eq. (3.19)) and form a complete set. The completeness property of Jacobi polynomials is [16]

$$\sum_{n=0}^{\infty} \left(h_n^{\alpha,\beta}\right)^{-1} P_n^{\alpha,\beta}(x) P_n^{\alpha,\beta}(y) = \frac{\delta(x-y)}{(1-x)^\alpha (1+x)^\beta}, \tag{3.56}$$

where $h_n^{\alpha\beta}$ is given by Eq. (3.20). Multiplying both sides of Eq. (3.55) by $(K'' + 1)! P_{K''}^{\frac{1}{2},\frac{1}{2}}(y)/\Gamma(K'' + \frac{3}{2})$, summing over K'' and using Eqs. (3.56), (3.20) and (3.26), we have

$$\sum_{K''=|K-K'|}^{K+K'} \frac{(K''+1)!}{\Gamma(K''+\frac{3}{2})} \left\langle {}^{(2)}P_{2K}^{l,l} \middle| {}^{(2)}P_{2K''}^{0,0} \middle| {}^{(2)}P_{2K'}^{l,l} \right\rangle P_{K''}^{\frac{1}{2},\frac{1}{2}}(y)$$

$$= N_{2K}^{l,l} N_{2K'}^{l,l} 2^{-(2l+1)} (1-y^2)^l P_{K-l}^{l+\frac{1}{2},l+\frac{1}{2}}(y) P_{K'-l}^{l+\frac{1}{2},l+\frac{1}{2}}(y). \tag{3.57}$$

The sum over K'' is restricted from $|K - K'|$ to $(K + K')$, which follows from Eq. (3.38). Equation (3.57) holds for any value of y in the interval $[-1, 1]$. Now for given values of K, K', l and y, Eq. (3.57) is a linear inhomogeneous equation (LIE) for the unknown 3-P coefficients. Since the 3-P coefficient vanishes unless K'' satisfies $|K - K'| \le K'' \le (K + K')$, there is a finite number $n_{K,K'} = (K + K') - |K - K'| + 1$ of unknowns. We can arbitrarily choose $n_{K,K'}$ different values of y in the interval $[-1, 1]$ and solve the set of $n_{K,K'}$ LIEs, Eq. (3.57), for the $n_{K,K'}$ unknown 3-P coefficients, for given values of K, K' and l. A computer code for solving a set of LIE is quite fast and very accurate. On the other hand, a direct numerical integration using Eq. (3.55) would be very slow and inaccurate [21, 22]. Note numerical integration has to be done for each of the $n_{K,K'}$ 3-P coefficients, while all of them will be obtained by solving the LIE only once.

A simple sum rule is given by setting $y = 1$ in Eq. (3.57) and using

$$P_n^{\alpha\beta}(1) = \binom{\alpha+n}{\alpha}$$

where

$$\binom{a}{b} = \frac{a!}{b!(a-b)!}$$

is the binomial coefficient. The result is (note that for $y = 1$, the right side of Eq. (3.57) vanishes for $l > 0$)

$$\sum_{K''=|K-K'|}^{K+K'} (K''+1) \left\langle {}^{(2)}P_{2K}^{l,l} \middle| {}^{(2)}P_{2K''}^{0,0} \middle| {}^{(2)}P_{2K'}^{l,l} \right\rangle = \delta_{l,0} \frac{4}{\sqrt{\pi}} (K+1)(K'+1). \tag{3.58}$$

Equation (3.58) gives a simple test for the correctness and accuracy of the calculated GSCs.

References

1. Fabre de la Ripelle, M.: The Hyperspherical Expansion Method—presented in 86th Autum School 'Models and Methods in Few-Body Physics', 13–18 October 1986, Lisbon (Portugal) published in 'Lecture Notes in Physics', vol. 273. Springer, Heidelberg (1987)
2. Krivec, R.: Few-Body Syst. **25**, 199 (1998)
3. Rosati, S.: The hyperspherical harmonics method: a review and some recent developments. In: Fabrocini, A., et al. (eds.) Introduction to Modern Methods of Quantum Many-Body Theory and Their Applications. World Scientific, Singapore (2002)
4. Zernike, F., Brinkman, H.C.: Proc. Kon. Ned. Akad. Wettensch. **38**, 161 (1935)
5. Delves, L. M.: Nucl. Phys. **9**, 391 (1959); **20**, 275 (1960)
6. Smith, F.T.: Phys. Rev. **120**, 1058 (1960); J. Math. Phys. **3**, 735 (1962)
7. Ballot, J.L., Fabre de la Ripelle, M.: Ann. Phys. (N.Y.) **127**, 62 (1980)
8. Avery, J.S.: Hyperspherical Harmonics: Applications in Quantum Theory. Springer, Netherland (1989)
9. Avery, J.S.: J. Phys. Chem. **97**, 2406 (1993); Avery, J.S.: J. Comp. Appl. Math. **233**, 1366 (2010)
10. Meremianin, A.V.: J. Math. Phys. **50**, 013526 (2009)
11. Wang, D., Kuppermann, A.: J. Phys. Chem. A **113**, 15384 (2009); Kuppermann, A.: Phys. Chem. Chem. Phys. **13**, 8259 (2011)
12. Avery, J.S.: Hyperspherical Harmonics and Generalized Sturmians. Springer, Berlin (2002)
13. Simonov, Yu.A.: Sov. J. Nucl. Phys. **3**, 461 (1966); Simonov, Yu. A.: Sov. J. Nucl. Phys. **7**, 722 (1968)
14. Vallieres, M., Coelho, H.T., Das, T.K.: Nucl. Phys. A **271**, 95 (1976); Coelho, H. T., Das, T.K., Vallieres, M.: Rev. Bras. Fis. **7**, 237 (1977)
15. Abramowitz, M., Stegun, I.A.: Handbook of Mathematical Functions. National Institute of Standards and Technology, USA (1964)
16. Arfken, G.: Mathematical Methods for Physicists. Academic Press, Waltham (1966)
17. Hosseinbor, A.P., et al.: 4D hyperspherical harmonic (HyperSPHARM) representation of multiple disconnected brain subcortical structures. In: Mori, K., et al. (eds.) MICCAI 2013, Part I. LNCS, vol. 8149, pp. 598–605. Springer, Berlin (2013)
18. Johnson, B.R.: J. Chem. Phys. **69**, 4678 (1978)
19. Raynal, J., Revai, J.: Nuovo Cimenti A **68**, 612 (1970)
20. Khan, Md.A., Dutta, S.K., Das, T.K.: Fizika B **8**, 469 (1999)
21. Das, T.K., De, T.B.: P.ramana J. Phys. **28**, 645 (1987)
22. De, T.B., Das, T.K.: Phys. Rev. C **36**, 402 (1987)

Chapter 4
General Many-Body Systems

Abstract Hyperspherical harmonics expansion method is generalized from three-body to A-body system, defining N Jacobi vectors and $3N$ hyperspherical variables for the relative motion, where $N = A - 1$. Analytic expression is derived for the generalized hyperspherical harmonics as the eigenfunction of the generalized hyperangular momentum operator. The kinematic rotation vector (KRV) is defined as a linear combination of the Jacobi vectors. Position vector of a particle from the center of mass and relative separation of a pair can be expressed as KRVs in terms of two sets of parametric angles. Procedure for symmetrization (including mixed symmetry) of the spatial wave function using KRV is described. It is emphasized that truncation of the basis is necessary for a practical calculation. Truncation schemes like restricting the symmetry component, retaining only the lowest hyperangular momentun (L_m approximation), restriction to optimal subset and the subset of potential harmonics have been introduced. Application of the truncation schemes to problems in particle, nuclear, and atomic physics has been discussed.

We discussed the hyperspherical harmonics method for the three-body system in the last chapter. However, one frequently encounters systems with larger number of particles in physics. Hence it is necessary to generalize the technique for a system containing an arbitrary number of particles. Hyperspherical techniques which are somewhat different from that presented in Chap. 3 have also been developed [1–4]. In this chapter, we will generalize the treatment presented in the previous chapter to a system of A identical spineless particles interacting through mutual forces, retaining the notations as close to those of Chap. 3 as possible. It will serve to illustrate how one can go gradually from one system to another with one more particle. We will start with the generalization of the Jacobi vectors and define the hyperspherical variables in Sect. 4.1. In the next section, the $3(A - 1)$-dimensional Laplace operator for the relative motion will be expressed in terms of the hyperspherical variables. In doing this, the definition of the hyperangular momentum operator and its eigenfunctions, the hyperspherical harmonics (HH) in this space will be introduced. In Sect. 4.3, we will discuss the expansion of the many-body wave function in the HH basis and give a general idea of how to impose the required symmetry on the wave function. Inclusion of this in the Schrödinger equation leading to a system of coupled differential

© Springer India 2016   33
T.K. Das, *Hyperspherical Harmonics Expansion Techniques*,
Theoretical and Mathematical Physics, DOI 10.1007/978-81-322-2361-0_4

equations will be discussed in Sect. 4.4. It will be seen that this procedure will be too complex to be used without truncation of basis or other simplifying approximations for $A > 3$. Various schemes of truncation will be discussed in Sect. 4.5, and their applications to physics will be presented in Sect. 4.6.

We have discussed time and again that for $A > 3$ we have to adopt approximations in order to keep the calculations manageable. The approximation stems basically from the restriction of the expansion basis of the wave function. These can be classified into two different types: first one is the restriction of the expansion basis to a subset of the full HH basis, arising from the interactions. For example, if the sum of all mutual interactions depends very weakly on the hyperangles, the grand orbital quantum number (K) becomes an approximately good quantum number and the subset of HH with a few lowest K values is sufficient. This is like the ground state of the deuteron having a major contribution from $\ell = 0$ only, since the nucleon–nucleon interaction is dominantly central. We will discuss a similar type of approximation, called the L_m approximation, and some of its applications in Sect. 4.5. Another very important example is the situation, in which the system is very dilute and only two-body correlations in the wave function are relevant. In this case one can restrict the expansion basis to a subset called '*potential harmonics*' (PH), which involves only the two-body correlations. The standard Bose–Einstein condensation achieved in the laboratory is a fine example of this approximation, which we will discuss in Chaps. 7 and 8.

The second type of approximation is to restrict the dominant symmetry components of the wave function. In general, a number of symmetry components (arising from the type of constituent particles and the conserved quantum numbers of the system) are allowed, but their relative importance in a particular state of the system may be dominant only for one or at most a few components. In that case, one can approximately restrict the wave function to the dominant symmetry components only. A general discussion can be found in Sect. 4.5.1. An application of the technique in connection with the trinucleon systems has been provided in Chap. 5.

4.1   Jacobi Coordinates and Hyperspherical Variables

We first ignore spin and isospin degrees of freedom and consider a system of A identical spinless particles of mass m each. In this case, there are $N = A - 1$ relative vectors and one CM vector. The relative vectors are not unique and we choose them as the Jacobi vectors, as a generalization of the three-body system, Eq. (3.2): the jth Jacobi vector being proportional to the vector separation of the $(j + 1)$th particle from the CM of the first j particles

$$\vec{\xi}_j = \left[\frac{2j}{j+1} \right]^{1/2} \left(\vec{r}_{j+1} - \frac{1}{j} \sum_{i=1}^{j} \vec{r}_i \right) \qquad (j = 1, \ldots, A - 1), \qquad (4.1)$$

where $\vec{r}_i$ is the position vector of the ith particle. Putting $j = 1$ and 2, we get back the two Jacobi vectors of the three-body system, Eq. (3.2). The CM vector is given by

$$\vec{R} = \frac{1}{A}\left(\sum_{i=1}^{A}\vec{r}_i\right). \tag{4.2}$$

The constants in front of the Jacobi vectors in Eq. (4.1) are chosen so that

$$\frac{1}{2}\sum_{i=1}^{A}\nabla_{\vec{r}_i}^2 = \sum_{j=1}^{A-1}\nabla_{\vec{\xi}_j}^2 + \frac{1}{2A}\nabla_{\vec{R}}^2. \tag{4.3}$$

Hence the A-body Schrödinger equation with mutual forces

$$\left[-\sum_{i=1}^{A}\frac{\hbar^2}{2m}\nabla_{\vec{r}_i}^2 + \sum_{j<i=2}^{A} V(\vec{r}_i - \vec{r}_j)\right]\psi(\vec{r}_1,\ldots,\vec{r}_A) = E_T\psi(\vec{r}_1,\ldots,\vec{r}_A) \tag{4.4}$$

separates into the Schrödinger equation for the $N = A - 1$ relative coordinates

$$\left[-\sum_{i=1}^{N}\frac{\hbar^2}{m}\nabla_{\vec{\xi}_i}^2 + V(\vec{\xi}_1,\ldots,\vec{\xi}_N)\right]\Psi(\vec{\xi}_1,\ldots,\vec{\xi}_N) = E\Psi(\vec{\xi}_1,\ldots,\vec{\xi}_N), \tag{4.5}$$

with relative energy E. The CM moves as a free particle with energy $E_T - E$, as in Eq. (3.4). Hence we have to study Eq. (4.5). As in Chap. 3, we will expand the relative wave function $\Psi(\vec{\xi}_1,\ldots,\vec{\xi}_N)$ in an appropriate basis, which will be the generalized hyperspherical harmonics basis.

Next we introduce the hyperspherical variables in the same manner as in Chap. 3. First define a 'hyperradius' through

$$\xi = \left[\sum_{j=1}^{N}\xi_i^2\right]^{1/2}. \tag{4.6}$$

In order to introduce the hyperangles, we follow the Zernike and Brinkman [5] representation, in which the $(3N - 1)$ 'hyperangles' are constituted by the $2N$ ordinary spherical polar angles of the N Jacobi vectors $[(\vartheta_i, \varphi_i), i = 1, \ldots, N]$ and $(N - 1)$ angles $(\phi_2,\ldots,\phi_N)$ defining the *relative lengths* of N Jacobi vectors $\vec{\xi}_1,\ldots,\vec{\xi}_N$, through

$\xi_N = \xi\cos\phi_N$
$\xi_{N-1} = \xi\sin\phi_N\cos\phi_{N-1}$
$\xi_{N-2} = \xi\sin\phi_N\sin\phi_{N-1}\cos\phi_{N-2}$

$$\xi_{N-3} = \xi \sin \phi_N \sin \phi_{N-1} \sin \phi_{N-2} \cos \phi_{N-3}$$

$$\vdots$$

$$\xi_2 = \xi \sin \phi_N \sin \phi_{N-1} \quad \cdots \quad \cdots \quad \cdots \quad \sin \phi_3 \cos \phi_2$$

$$\xi_1 = \xi \sin \phi_N \sin \phi_{N-1} \quad \cdots \quad \cdots \quad \cdots \quad \sin \phi_3 \sin \phi_2 \cos \phi_1, \qquad (4.7)$$

with $\phi_1 = 0$. This definition is consistent with Eq. (3.6) for $N = 2$. Each ϕ_i lies in the interval $[0, \pi/2]$. Equation (4.7) satisfies Eq. (4.6) automatically. In analogy with Chap. 3, we denote the collection of these $(3N - 1)$ hyperangles by $\Omega_{3N} = \{(\vartheta_1, \varphi_1), \ldots, (\vartheta_N, \varphi_N), \phi_2, \ldots, \phi_N\}$. This choice of hyperangles has the advantage that the angular momentum $\vec{l}_j$ carried by the jth Jacobi vector $\vec{\xi}_j$ appears naturally and the total orbital angular momentum $\vec{l}$ of the relative motion is simply the vector sum of the N individual angular momenta: $\vec{l} = \vec{l}_1 + \vec{l}_2 + \cdots + \vec{l}_N$. However, there is a serious disadvantage that the separation vector of two particles $\vec{r}_{ij} = \vec{r}_j - \vec{r}_i$ becomes a very complicated function of the hyperangles for arbitrary values of i and j. Only in a special case, it is simple. For example, from Eq. (4.1), we see that $\vec{r}_{12} = \vec{\xi}_1$. This is particularly important in handling the mutual potential. An alternative to this complexity is to use an appropriately symmetrized basis, so that the matrix element of the total mutual interaction becomes simply the number of pairs multiplied by the matrix element of any one pair interaction. Since this pair can be chosen arbitrarily, it can be chosen as the (12)-pair. However, the complexity of the calculation goes over to the calculation of the appropriately symmetrized basis. Indeed there is no simple way to avoid the complexity for more than three interacting particles.

4.2 Generalized Hyperspherical Harmonics

Next, we express the $(3N)$-dimensional Laplace operator

$$\nabla^2_{3N} = \sum_{j=1}^{N} \nabla^2_{\vec{\xi}_j}$$

in terms of the chosen hyperspherical variables $\{\xi, \Omega_{3N}\}$ [6] (we continue with the choice $\hbar = 1$)

$$\sum_{j=1}^{N} \nabla^2_{\vec{\xi}_j} = \left(\frac{\partial^2}{\partial \xi^2} + \frac{3N - 1}{\xi} \frac{\partial}{\partial \xi} - \frac{\hat{\mathcal{L}}^2_{3N}(\Omega_{3N})}{\xi^2} \right), \qquad (4.8)$$

where

$$\hat{\mathcal{L}}_{3N}^2(\Omega_{3N}) = -\sum_{i=1}^{N} \left(\prod_{j=i+1}^{N} \sin^2 \phi_j \right)^{-1}$$

$$\times \left[\frac{\partial^2}{\partial \phi_i^2} + ((3i-4)\cot\phi_i - 2\tan\phi_i)\frac{\partial}{\partial\phi_i} - \frac{\hat{l}_i^2(\vartheta_i, \varphi_i)}{\cos^2\phi_i} \right].$$

(4.9)

Note that the factor in front of the third brackets in Eq. (4.9) takes the value 1 for $i = N$. We can easily verify that Eqs. (4.8) and (4.9) give Eqs. (3.7) and (3.8), respectively, for $N = 2$.

The generalized hyperspherical harmonics $\mathcal{Y}_{[L]}(\Omega_{3N})$ is defined in analogy with Chap. 3, as the eigenfunction of $\hat{\mathcal{L}}_{3N}^2(\Omega_{3N})$, satisfying the Laplace equation in $(3N)$ dimensional space

$$\sum_{j=1}^{N} \nabla^2_{\xi_j} \left\{ \xi^L \mathcal{Y}_{[L]}(\Omega_{3N}) \right\} = 0. \tag{4.10}$$

Thus $\{\xi^L \mathcal{Y}_{[L]}(\Omega_{3N})\}$ is a homogeneous harmonic polynomial of order L in the Cartesian components of N Jacobi vectors. Using Eqs. (4.8) and (4.10), the eigenvalue equation satisfied by the hyperspherical harmonics (HH) becomes

$$\hat{\mathcal{L}}_{3N}^2(\Omega_{3N})\mathcal{Y}_{[L]}(\Omega_{3N}) = L(L+3N-2)\mathcal{Y}_{[L]}(\Omega_{3N}). \tag{4.11}$$

L is called the *hyperangular momentum* (or grand orbital) quantum number and the symbol $[L]$ denotes the full set of $(3N-1)$ quantum numbers for a fixed L. These quantum numbers in the chosen Zernike–Brinkman representation are

$$[L] \equiv \{L; (l_1, m_1), \ldots, (l_N, m_N), n_2, n_3, \ldots, n_N\}, \tag{4.12}$$

where the nonnegative integer n_j is the quantum number associated with the hyperangular variable ϕ_j. Note that L is given in terms of l_j and n_j by

$$L = \sum_{j=1}^{N}(2n_j + l_j) \qquad (n_1 = 0). \tag{4.13}$$

Thus the number of independent quantum numbers in Eq. (4.12) is $(3N - 1)$.

The form of Eq. (4.9) shows that each (ϑ_i, φ_i) separates and the corresponding function is the standard spherical harmonics (an eigenfunction of $\hat{l}_i^2$). Also each of ϕ_i separates. Hence the HH will be a product of spherical harmonics $Y_{l_i m_i}(\vartheta_i, \varphi_i)$ and functions of ϕ_i. A detailed calculation similar to the three-body case gives [6]

$$\mathcal{Y}_{[L]}(\Omega_{3N}) = Y_{l_1 m_1}(\vartheta_1, \varphi_1) \prod_{j=2}^{N} Y_{l_j m_j}(\vartheta_j, \varphi_j) \,\, {}^{(j)}\mathcal{P}_{L_j}^{l_j, L_{j-1}}(\phi_j), \tag{4.14}$$

where ${}^{(j)}\mathcal{P}_{L_j}^{l_j, L_{j-1}}(\phi_j)$ is given by

$$
{}^{(j)}\mathcal{P}_{L_j}^{l_j, L_{j-1}}(\phi_j) = \left[\frac{2\nu_j \Gamma(\nu_j - n_j)\Gamma(n_j + 1)}{\Gamma(\nu_j - n_j - l_j - \frac{1}{2})\Gamma(n_j + l_j + \frac{3}{2})} \right]^{1/2}
$$
$$
\times (\cos \phi_j)^{l_j} (\sin \phi_j)^{L_{j-1}} P_{n_j}^{\nu_{j-1}, l_j + \frac{1}{2}}(\cos 2\phi_j) \quad (j \geq 2). \tag{4.15}
$$

Here, $P_n^{\alpha, \beta}(x)$ is a Jacobi polynomial [7] and the quantities ν_j and L_j are given as

$$
L_j = \sum_{i=1}^{j}(2n_i + l_i) \qquad (n_1 = 0)
$$
$$
\nu_j = L_j + \frac{3j}{2} - 1. \tag{4.16}
$$

It can be verified that these expressions reduce to the HH for the three-body system for $N = 2$.

Under ordinary parity operation each Jacobi vector changes sign, the length remaining unchanged. Hence all the ϕ_j remain unchanged. From Eq. (4.14), we see that the effect of parity operation introduces a factor

$$(-1)^{(l_1 + l_2 + \cdots + l_N)} = (-1)^{[L - 2(n_2 + n_3 + \cdots + n_N)]} = (-1)^L$$

Thus the parity of the HH is given by the grand orbital L.

The A-body relative wave function can be expanded in the complete set of HH given by Eqs. (4.14) and (4.15)

$$\Psi(\vec{\xi}_1, \ldots, \vec{\xi}_N) = \xi^{-(3N-1)/2} \sum_{L, [L]} u_{[L]}(\xi) \mathcal{Y}_{[L]}(\Omega_{3N}). \tag{4.17}$$

The factor $\xi^{-(3N-1)/2}$ in front is included to remove the first derivative term in the coupled differential equations resulting from the Schrödinger equation. The sum over L, $[L]$ in Eq. (4.17) means a sum over L from 0 to ∞ and then for a particular value of L, the sum over $[L]$ indicates the sum over the $(3N - 1)$ quantum numbers indicated in Eq. (4.12).

4.3 Symmetrization of Wave Function

4.3.1 Kinematic Rotation Vector (KRV)

Although nonsymmetrized hyperspherical harmonics have been used [8], it is customary to use appropriately symmetrized basis. For symmetrization, we need different permutations of the particle indices. It is clear from the definition of Jacobi vectors Eq. (4.1) that in general a permutation of the indices will change a given Jacobi vector into a linear combination (l.c.) of the original set. Also the separation vector $\vec{r}_{ij}$, as also the position of the ith particle from the CM are l.c.'s of the original Jacobi vectors. We introduce a 3D vector, called *kinematic rotation vector*, which is the most general l.c. of Jacobi vectors $\{\vec{\xi}_1, \ldots, \vec{\xi}_N\}$ by [9]

$$\vec{z}(\varpi) = \sum_{j=1}^{N} \sin \varpi_N \, \sin \varpi_{N-1} \, \ldots \, \sin \varpi_{j+1} \, \cos \varpi_j \, \vec{\xi}_j, \qquad (4.18)$$

where the set of $(N-1)$ angles, $\{\varpi_1 \equiv 0, \varpi_2, \ldots, \varpi_N\}$, define the particular l.c. Note the similarity in the way the coefficients in terms of the angular parameters are chosen with the introduction of hyperangles in Eq. (4.7). In both cases, the sum of the squares of the coefficients is one. The advantage of this choice will be seen in the following. Consider an arbitrary l.c. of the Jacobi vectors

$$\vec{A}(\varpi) = \sum_{j=1}^{N} a_j \vec{\xi}_j. \qquad (4.19)$$

Note that in the above the argument (ϖ) of a 3D vector is a short hand for the full set of these angles. In order to express $\vec{A}(\varpi)$ in terms of $\vec{z}(\varpi)$, we can take the coefficient a_j to be proportional to the coefficient of $\vec{\xi}_j$ in Eq. (4.18). However, the sum of squares of the latter coefficients is identically one. So we write

$$\vec{A}(\varpi) = C\vec{z}(\varpi). \qquad (4.20)$$

It is easy to find that

$$\cos^2 \varpi_j = \frac{a_j^2}{\sum_{i=1}^{j} a_i^2}, \qquad (4.21)$$

and

$$C^2 = \sum_{j=1}^{N} a_j^2. \qquad (4.22)$$

An advantage of the particular choice of the parametric angles in Eq. (4.18) is that it gives rise to a simple expression for the parametric angle in Eq. (4.21).

Two sets of these parametric angles with superscripts (i) and (ij) will be used to denote the position of the ith particle from the CM and the (ij)-pair separation vector, respectively,

$$\vec{z}(\varpi^{(i)}) = \sqrt{\frac{2(N+1)}{N}}(\vec{r}_i - \vec{R})$$
$$\vec{z}(\varpi^{(i,j)}) = \vec{r}_j - \vec{r}_i. \tag{4.23}$$

Coefficients of KRV for Pair Separation

(a) Consecutively numbered particles

In order to calculate the coefficients of expansion of Eq. (4.23) in the original set of Jacobi vectors, we first express the relative separation vector of two consecutively numbered particles, using Eq. (4.1)

$$\vec{r}_{i+1,i} = \vec{r}_{i+1} - \vec{r}_i = -\sqrt{\frac{i-1}{2i}}\vec{\xi}_{i-1} + \sqrt{\frac{i+1}{2i}}\vec{\xi}_i. \tag{4.24}$$

Thus, $\vec{r}_{i+1,i}$ contains only two Jacobi vectors, viz., $\vec{\xi}_{i-1}$ and $\vec{\xi}_i$. Following Eq. (4.19), we write

$$\vec{r}_{i+1,i} = \sum_{j=1}^{N} a_j^{(i+1,i)}\vec{\xi}_j. \tag{4.25}$$

Then comparing Eqs. (4.24) and (4.25), the coefficients $a_j^{(i+1,i)}$ are given as

$$\begin{aligned} a_j^{(i+1,i)} &= 0 & j < (i-1) \\ &= 0 & j > i, \end{aligned}$$

and

$$a_{i-1}^{(i+1,i)} = -\sqrt{\frac{i-1}{2i}}$$
$$a_i^{(i+1,i)} = \sqrt{\frac{i+1}{2i}}. \tag{4.26}$$

In Table 4.1, we display the coefficients $a_j^{(i+1,i)}$ for $j = 1, N$ (along the columns) of consecutively numbered $(i+1, i)$-pair separation with $i = 1, N$ (along the rows).

(b) Arbitrary pair separation

For an arbitrary $\vec{r}_{i,j}$ with $i > j$, we can write

$$\vec{r}_{i,j} = \vec{r}_{i,i-1} + \vec{r}_{i-1,i-2} + \cdots + \vec{r}_{j+1,j}, \tag{4.27}$$

Table 4.1 $a_j^{(i+1,i)}$ coefficients for expansion of $\vec{r}_{i+1,i}$ according to Eq. (4.19)

$(i+1,i)$ \ j	1	2	3	4	...	$N-3$	$N-2$	$N-1$	N
$(2,1)$	1	0	0	0	...	0	0	0	0
$(3,2)$	$-\frac{1}{2}$	$\frac{\sqrt{3}}{2}$	0	0	...	0	0	0	0
$(4,3)$	0	$-\frac{1}{\sqrt{3}}$	$\sqrt{\frac{2}{3}}$	0	...	0	0	0	0
$(5,4)$	0	0	$-\sqrt{\frac{3}{8}}$	$\sqrt{\frac{5}{8}}$	...	0	0	0	0
...					...				
$(N-1,N-2)$	0	0	0	0	...	$-\sqrt{\frac{N-3}{2(N-2)}}$	$\sqrt{\frac{N-1}{2(N-2)}}$	0	0
$(N,N-1)$	0	0	0	0	...	0	$-\sqrt{\frac{N-2}{2(N-1)}}$	$\sqrt{\frac{N}{2(N-1)}}$	0
$(N+1,N)$	0	0	0	0	...	0	0	$-\sqrt{\frac{N-1}{2N}}$	$\sqrt{\frac{N+1}{2N}}$

and use Eq. (4.26) repeatedly. As an example, let us calculate $\vec{r}_{5,2}$. We write it following Eq. (4.27) and then for each consecutively numbered pair, use Table 4.1

$$\vec{r}_{5,2} = \vec{r}_{5,4} + \vec{r}_{4,3} + \vec{r}_{3,2}$$

$$= \left(-\sqrt{\frac{3}{8}}\vec{\xi}_3 + \sqrt{\frac{5}{8}}\vec{\xi}_4\right) + \left(-\frac{1}{\sqrt{3}}\vec{\xi}_2 + \sqrt{\frac{2}{3}}\vec{\xi}_3\right) + \left(-\frac{1}{2}\vec{\xi}_1 + \frac{\sqrt{3}}{2}\vec{\xi}_2\right)$$

$$= -\frac{1}{2}\vec{\xi}_1 + \frac{1}{2\sqrt{3}}\vec{\xi}_2 + \frac{1}{\sqrt{24}}\vec{\xi}_3 + \sqrt{\frac{5}{8}}\vec{\xi}_4. \tag{4.28}$$

Thus we get $\vec{r}_{5,2}$ in terms of the original set of Jacobi vectors. A look at Table 4.1 shows that for an arbitrary $\vec{r}_{i,j}$ with $i > j$ written as Eq. (4.27), one simply adds the numbers in the kth column from the $(j+1, j)$ row to $(i, i-1)$ row, and multiply by $\vec{\xi}_k$. Finally add all the nonvanishing contributions, $k = j-1$ to $k = i-1$. The pair separation vectors are useful in calculating the potential matrix.

Coefficients of KRV for Pair Exchange Operations

Under P_{ij}, which exchanges particle labels of the pair (ij), the new set of Jacobi vectors becomes a l.c. of the original set. For example, we have from Eq. (4.1)

$$P_{12}\vec{\xi}_1 = -\vec{\xi}_1$$
$$P_{12}\vec{\xi}_k = \vec{\xi}_k \qquad k \geq 2, \tag{4.29}$$

and

$$P_{23}\vec{\xi}_1 = \frac{1}{2}\vec{\xi}_1 + \frac{2}{\sqrt{3}}\vec{\xi}_2$$

$$P_{23}\vec{\xi}_2 = \frac{3\sqrt{3}}{8}\vec{\xi}_1 - \frac{1}{2}\vec{\xi}_2$$

$$P_{23}\vec{\xi}_k = \vec{\xi}_k \qquad k \geq 3. \tag{4.30}$$

We can easily verify by applying P_{ij} twice that $(P_{ij})^2 = 1$. One can construct a table of the expansion coefficients in a manner similar to Table 4.1. Pair exchange operators can be used for symmetrization of the wave function since any permutation of the particle indices can be written as a product of pair exchange operators.

Calculation of the Parametric Angles $\varpi^{(i,j)}$ for the pair separation

The coefficients given by Eq. (4.26) will satisfy Eq. (4.18), if we choose the last cosine factor of each term with $j > i$ to be zero and the last cosine factor of each term with $j < i$ to be one (so that the corresponding sine is zero), while $\cos \varpi_i^{(i+1,i)}$ and $\sin \varpi_i^{(i+1,i)}$ are given by the last two entries of Eq. (4.26)

$$\varpi_j^{(i+1,i)} = \pi/2 \qquad\qquad j > i$$
$$= 0 \qquad\qquad j < i,$$

and

$$\cos \varpi_i^{(i+1,i)} = -\sqrt{\frac{i-1}{2i}}$$
$$\sin \varpi_i^{(i+1,i)} = \sqrt{\frac{i+1}{2i}}. \tag{4.31}$$

Thus we get the parametric angles to express the separation vector of two consecutively numbered particles, using Eqs. (4.18) and (4.31). Fabre gives a tabular presentation of these parametric angles in Ref [9]. However, note that the definition of the original set of Jacobi vectors in Ref. [9] is in the reverse order of our choice Eq. (4.1), such that our $\vec{\xi}_i$ is $\vec{\xi}_{N-i+1}$ used in Ref. [9].

4.3.2 Symmetrization of Wave Function

In a fashion similar to Eq. (4.23), we can introduce the set of parametric angles $(\varpi^{(i,P)})$ for the ith Jacobi vector $(\vec{\xi}_i^{(P)})$ under a particular permutation (P) of the particle indices. Using this, one can construct a desired symmetry component of the space wave function, as in Chap. 3. For example, the totally symmetric space wave function is

$$\Psi^{(S)}(\vec{\xi}_1, \ldots, \vec{\xi}_N) = \sum_P \Psi^{(P)}(\vec{\xi}_1^{(P)}(\varpi^{(1,P)}), \ldots, \vec{\xi}_N^{(P)}(\varpi^{(N,P)}))$$
$$= \xi^{-(3N-1)/2} \sum_{L,[L]} u_{[L]}(\xi) \sum_P \mathcal{Y}_{[L]}(\Omega_{3N}^{(P)}), \tag{4.32}$$

where the hyperangles $\Omega_{3N}^{(P)}$ is obtained for the set of Jacobi vectors for the permutation (P), viz., $(\vec{\xi}_1^{(P)}(\varpi^{(1,P)}), \ldots, \vec{\xi}_N^{(P)}(\varpi^{(N,P)}))$. Note that the hyperradius ξ is invariant under all permutations and under all 3D rotations. We can calculate the effects of permutations, i.e., the set of angles $\{\varpi^{(i,P)}, i = 1, \ldots, N\}$, for all $(N + 1)!$ permutations (P), calculate the HH explicitly and use Eq. (4.32). However, this is a very laborious process for $N > 3$. Alternatively, we can expand $\mathcal{Y}_{[L]}(\vec{\xi}_1^{(P)}(\varpi^{(1,P)}), \ldots, \vec{\xi}_N^{(P)}(\varpi^{(N,P)}))$ in the complete set of HH of the original partition, defining coefficients similar to Raynal–Revai coefficients for the three-body system. This again will be too cumbrous to follow for more than three particles.

However luckily for most systems of interest in physics and chemistry, only a small subset of the full HH set contributes significantly in the expansion of the wave function. We will see in Sect. 4.6 that the symmetrization may become more manageable for the relevant subset of HH. Use of the subset is an approximation which is guided by the physics of the system.

We next consider inclusion of different symmetry components of the total wave function (due to spin and isospin, where the latter is relevant). Although use of the complete set of HH is not very practical, the procedure is the same for a subset. A particular subset may contain a smaller number of symmetry components. For the expansion of the total wave function (having total angular momentum J and its projection M_J) we replace Eq. (4.17) by

$$
\Psi_{JM_J}(\vec{\xi}_1, \ldots, \vec{\xi}_N, \vec{s}_1, \ldots, \vec{s}_A, \vec{t}_1, \ldots, \vec{t}_A) =
$$
$$
\xi^{-(3N-1)/2} \sum_{S} \sum_{L,[L]} u_{[L]}^{(S)}(\xi) \left[\mathcal{Y}_{[L]}^{(S)}(\Omega_{3N}) \chi^{(S)}(\vec{s}_1, \ldots, \vec{s}_A, \vec{t}_1, \ldots, \vec{t}_A) \right]_{JM_J}, \quad (4.33)
$$

where the superscript (S) refers to a particular symmetry component and $\chi^{(S)}$ is the spin-isospin wave function corresponding to the symmetry component (S). The symbol $[\cdots]_{JM_J}$ represents angular momentum coupling of total orbital $(\vec{l})$ and total spin $(\vec{s})$ angular momenta. The symmetries of $\mathcal{Y}^{(S)}$ and $\chi^{(S)}$ are conjugate to each other, such that their product has the desired symmetry of the system. For example, if $\mathcal{Y}^{(S)}$ is totally symmetric, then $\chi^{(S)}$ is totally antisymmetric, for a system of A identical fermions. Appropriate symmetrization of $\mathcal{Y}^{(S)}$ can be done using the results of Sect. 4.3.1. The symmetrization of $\chi^{(S)}$ is done using angular momentum algebra and selecting appropriate angular momentum coupling [10, 11]. We will see an explicit example for the trinucleon systems in the next chapter.

4.4 Schrödinger Equation: Coupled Differential Equations

Substitution of the expansion, Eq. (4.33), in the Schrödinger equation for the relative motion, Eq. (4.5), premultiplication by $\mathcal{Y}_{[L]}^{(S)*}(\Omega_{3N})$, and integration over $d\Omega_{3N}$ give rise to a system of coupled differential equations (CDE)

$$\left(-\frac{\hbar^2}{m}\frac{d^2}{d\xi^2} + \frac{\mathcal{L}(\mathcal{L}+1)}{\xi^2} - E\right)u_{[L]}^{(S)}(\xi)$$

$$+\sum_{S'}\sum_{L',[L']}\left\langle\left[\mathcal{Y}_{[L]}^{(S)}(\Omega_{3N})\chi^{(S)}(\vec{s}_1,\ldots,\vec{s}_A,\vec{t}_1,\ldots,\vec{t}_A)\right]_{JM_J}\right| V$$

$$\left|\left[\mathcal{Y}_{[L']}^{(S')}(\Omega_{3N})\chi^{(S')}(\vec{s}_1,\ldots,\vec{s}_A,\vec{t}_1,\ldots,\vec{t}_A)\right]_{JM_J}\right\rangle u_{[L']}^{(S')}(\xi) = 0, \qquad (4.34)$$

where $\mathcal{L} = L + (3N - 3)/2$. In the derivation of the above equation, Eqs. (4.8) and (4.11) have been used. Note that the sets $\{\mathcal{Y}_{[L]}^{(S)}(\Omega_{3N})\}$ and $\{\chi^{(S)}\}$ are orthonormal ones. Note also that both the hyperradius ξ and its associated quantum number (the hyperangular momentum or 'grand orbital' quantum number) L remain invariant under all permutations. But in general L is not a good quantum number of the system. In Eq. (4.34), V is the total interaction of all particles, expressed in terms of the Jacobi vectors. Following Sect. 4.3.1, all mutual separations can be expressed in terms of the original set of Jacobi vectors. The potential V can, in general, involve spin and isospin operators, which act on $\chi^{(S)}$. The coupling matrix element of Eq. (4.34) can be evaluated directly. Alternately, one can expand the potential in an appropriate set of HH. The expansion coefficients become the multipoles, which are functions of ξ. Then, as in Chap. 3, we can calculate the potential matrix element as a sum of products of potential multipoles and geometrical structure coefficients.

The sum over L in Eq. (4.34) is an infinite one. For a numerical calculation, the upper limit has to be restricted. This corresponds to a truncation of the full HH basis. For most physical systems with a realistic interaction, the contribution to the expansion (4.33) decreases rapidly with increase in L. This is so because for large L the partial wave will be peaked at a much larger value of ξ, while the low-lying bound states of the system will be highly localized. The accuracy of such a truncation is controllable numerically and if no other approximation is made, such a calculation is referred to as an *essentially exact* calculation.

In addition to the fact that the procedure becomes very involved for larger N, it also depends on the nature of interaction and the number and nature of symmetry components, the conserved quantum numbers, etc. All these may be much simplified in an approximate way by the choice of a suitable subset of the full HH basis. Thus, it is not very convenient to discuss the procedure in a general way. In the next section, we will discuss approximations due to the use of subsets. In Chap. 5, applications of the technique will be presented.

4.5   Approximation by Truncation of Basis

If the pairwise interaction is a purely central one, independent of spin and isospin, the full set of HH can be retained, with only an upper limit in the L sum in Eq. (4.34). Coulomb systems like two-electron atoms, muonic atoms, *etc.* fall in this category and will be discussed in Chap. 6. Effect of the straight forward truncation of the full basis (keeping up to a maximum of L value of L_{max}) for two-electron atoms was investigated by Chattopadhyay and Das [12]. They found that the binding energy converges smoothly with L_{max} and can be extrapolated for $L_{max} \to \infty$. But such essentially exact calculations are not possible for few-quark and few-nucleon systems, for which the interaction is noncentral and depends on spin and isospin variables. Even for the simplest nontrivial system, *viz.*, the trinucleon, there are several symmetry components and the calculations become very involved, which will be discussed in Chap. 5. Thus approximation schemes are needed. A common type involves truncation of the HH basis to be chosen for the expansion of the wave function. We discuss some truncation schemes in the following.

4.5.1 Restriction of Symmetry Components

The simplest approximation is to restrict the symmetry components of the space wave function arising from different symmetries under exchange of spin and isospin of three or more identical particles. For the good quantum number J of a selected state, different combinations of l and s may contribute. For example, for the ground states ($J = \frac{1}{2}$) of trinucleon systems (^{3}H and ^{3}He nuclei), there are three major components, namely the space totally symmetric $l = 0$ state (S state), space mixed symmetry $l = 0$ state (S' state), and the $l = 2$ state (D state). Out of these the S state contributes almost 90 % to the ground state and as a first approximation, the ground states of trinucleon can be taken as the space totally symmetric S state. This will be discussed in detail in Chap. 5.

Under this scheme, one can also truncate the expansion basis to a subset that retains the most important and dominant features of the system and the interactions. The justification of this truncation follows from the physics of the system. For example, if the tensor interaction is not very dominant, it can be disregarded, so that l becomes a good quantum number and the expansion basis for the ground state can be restricted to the subset corresponding to $l = 0$ only.

4.5.2 L_m Approximation

As in the case of the three-body system (Chap. 3 Sect. 3.4.1), the total potential $V(\vec{\xi}_1, \ldots, \vec{\xi}_N)$ can be expanded in the HH basis

$$V(\vec{\xi}_1, \ldots, \vec{\xi}_N) = \sum_{L,[L]} V_{[L]}(\xi) \mathcal{Y}_{[L]}(\Omega_{3N}). \tag{4.35}$$

If the potential contains spin and isospin operators, the potential multipole $V_{[L]}(\xi)$ will involve the same. If the potential is hypercentral, i.e., independent of Ω_{3N}, then only the $L = 0$ term survives. In this case both l and L become good quantum numbers and the CDE (4.34) reduces to a single uncoupled differential equation. For the ground state both these take zero eigen values. The wave function Ψ_0 becomes hyperspherically symmetric (independent of hyper angles). For other states or if the potential is not hypercentral, the lowest L may have a nonvanishing value L_m, consistent with symmetry and conserved angular momenta. Since the hypercentrifugal repulsion of Eq. (4.34) is repulsive and increases rapidly with $\mathcal{L}$, the ground state will have dominant contribution from L_m. A purely hypercentral potential is unlikely in a realistic situation. Even if the two-body potential is purely central, $V(\vec{\xi}_1, \ldots, \vec{\xi}_N)$ depends on the hyperangles (except for the case of harmonic oscillator potential). In this case, Eq. (4.35) will have contributions from higher order multipoles and the wave function Ψ will deviate from hyperspherical symmetry. If the dependence of the total interaction on the hyperangles is weak compared to the hypercentral part, one can restrict the expansion of the wave function to a single term $\mathcal{Y}_{[L_m]}(\Omega_{3N})$. This is referred to as the L_m approximation.

4.5.3 Optimal Subset

On the other hand, if a better approximation is desired, one can use the optimal subset. When the hypercentral part of the total potential is the dominant one, their difference can be treated as a perturbation. Consider the subset of HH which is directly connected to the lowest dominant member $\mathcal{Y}_{[L_m]}(\Omega_{3N})$ through the potential $V(\vec{\xi}_1, \ldots, \vec{\xi}_N)$, i.e., the subset of HH $\{\mathcal{Y}_{[L]}\}$ should satisfy

$$\langle \mathcal{Y}_{[L]} | V | \mathcal{Y}_{[L_m]} \rangle = \int \mathcal{Y}^*_{[L]}(\Omega_{3N}) V(\vec{\xi}_1, \ldots, \vec{\xi}_N) \mathcal{Y}_{[L_m]}(\Omega_{3N}) d\Omega_{3N} \neq 0. \quad (4.36)$$

This subset is called the 'optimal subset'. Clearly, it depends on the nature of the potential and the most dominant HH $\mathcal{Y}_{[L_m]}(\Omega_{3N})$. Use of the optimal subset corresponds to a perturbative calculation up to and including the third order [6]. The use of the optimal subset for the trinucleon system will be taken up in Chap. 5, Sect. 5.3.

4.5.4 Potential Harmonics

The potential harmonics (PH) is the subset of HH which is sufficient for the expansion of the two-body potential $V(\vec{r}_{ij})$. Clearly, it depends on the label (ij) of the pair. If the system of particles interacts through two-body interactions alone, the wave function can be decomposed in Faddeev components of all interacting pairs. The (ij) Faddeev component (which represents the (ij)-pair interacting, while all the remaining particles are inert spectators), in general, depends on position vectors of all the particles. However, if the system is very dilute, correlations higher than two-body ones can be disregarded and the (ij) Faddeev component becomes a function of $\vec{r}_{ij}$ and hyperradius only. Then it can be expanded in the PH subset corresponding to (ij)-pair. Such an expansion retains only two-body correlations in the spatial wave function.

Correlations in the Wave Function

The full HH set takes into account all correlations and hence is necessarily a larger set than the PH. If the wave function depends on the position of each particle, but is independent of the relative separations (this is the case when the total wave function is a product of single particle wave functions, the ith member of which depends on the position of the ith particle only), then the total wave function has no correlations. This means that the probability of finding the ith particle at $\vec{r}_i$ is independent of the positions of all remaining particles. On the other hand, the total wave function has n-body correlation if the probability of finding the ith particle at $\vec{r}_i$ depends on the relative positions of $(n-1)$ other particles in its neighborhood, in addition to its own position.

Consider a few-body system which interacts through two-body forces only. If the system is very dilute such that the average interparticle separation is much larger than the range of interaction (R_0), then the probability that more than two particles will be simultaneously within a sphere of radius R_0 will be negligible. Hence the wave function will depend on relative separations of pairs of particles only. In this case, contributions of correlations higher than two-body ones in the wave function is negligible. For such a dilute system the PH subset is sufficient for the expansion of the spatial wave function of the system. This simplifies the analytic and numerical calculations greatly. An example is the dilute Bose–Einstein condensate (BEC), which is a cloud of bosonic atoms below the critical temperature. The number density of the cloud must be extremely low, so that there is no appreciable depletion due to formation of molecules and clusters through three-body and higher body collisions. We will discuss the potential harmonics expansion in Chap. 7 and its application to BEC in Chap. 8.

In passing, we discuss how three or higher body correlations may appear with two-body forces only. Consider a denser system having a finite probability of finding more than two particles simultaneously within the sphere of influence of radius R_0. Even with only two-body forces, the wave function of the system will depend on the relative configuration of all the particles within this sphere and thus many-body correlations enter into the picture. Higher body correlations become relevant as the system becomes denser. Clearly, the PH subset is not adequate for such a system.

4.6 Truncation of Basis: Application to Particles and Nuclei

The hyperspherical harmonics expansion method (HHEM) with truncation of the expansion basis according to one or more of the schemes given in the last section will be briefly discussed in this section. Since we present these as examples of the technique, discussions of the details of calculations and their results are outside the scope of this section. Interested readers can find the details from the references provided.

Few-body systems that have been studied by hyperspherical harmonics expansion method include baryons treated as three-quark systems, three and four nucleon systems, clusters (trimer, tetramer, etc.) of inert gases, etc. In the following subsections, we present a brief description of some of these as examples of truncation of the HH basis. Because the trinucleon has a special importance in physics, as also historically it was the first system studied by the HHEM in detail, we reserve it for a more detailed description in Chap. 5. Furthermore, for its importance in dilute many-body systems, the potential harmonics will be discussed in detail in Chap. 7 and its application to the Bose–Einstein condensates will be presented in Chap. 8.

4.6.1 Baryons as Three-Quark Systems

Spectra and other properties of nucleons, Δ resonances, and strange hyperons have been described by constituent three-quark model [13–19]. In this case, there are additional degrees of freedom like color and flavor. The symmetry components were restricted as follows. Requirement of color singlet states demands that the color wave function be antisymmetric. Overall antisymmetry means that space-spin-flavor wave function be totally symmetric under any pair exchange. Thus individual mixed symmetry space and spin-flavor states will contribute. A truncation scheme is adopted by treating the total wave function as a product of the space totally symmetric wave function (S), symmetric spin-flavor wave function, and the antisymmetric color wave function. The mixed symmetry space contributions are disregarded by the approximation scheme of Sect. 4.5.1. From our earlier discussion of Sect. 4.5.2 we know that the lowest grand orbital quantum number contributes most to the ground state. Hence, as a further simplification, the grand orbital quantum number is restricted to the lowest value, by the L_m approximation, to reduce the set of CDE to a single differential equation.

The nucleon wave function is thus given by

$$\psi^N = \frac{1}{\sqrt{2}}(\chi^\rho\eta^\rho + \chi^\lambda\eta^\lambda)u_N(\xi)\xi^{-5/2}, \qquad (4.37)$$

where $\chi^\rho(\eta^\rho)$ are the mixed antisymmetric spin (flavor) wave functions and $\chi^\lambda(\eta^\lambda)$ are the mixed symmetric spin (flavor) wave functions. The wave function for Δ is given by

$$\psi^\Delta = \chi^{\frac{3}{2}}\eta^{\frac{3}{2}}u_\Delta(\xi)\xi^{-5/2}. \qquad (4.38)$$

The hyperradial wave functions are obtained by solving single differential equations

$$\left[\frac{\hbar^2}{m}\left(-\frac{d^2}{d\xi^2}+\frac{15/4}{\xi^2}\right)+V_{N,\Delta}-E\right]u_{N,\Delta}(\xi)=0, \tag{4.39}$$

where $V_{N,\Delta}(\xi)$ are the single potential matrix element corresponding to L_m term and is given by

$$V_{N,\Delta}(\xi)=\frac{48}{\pi}\int_0^1\left[V^0(\xi u)-V^S(\xi u)+C_{N,\Delta}V^X(\xi u)\right]u^2\sqrt{1-u^2}\,du, \tag{4.40}$$

where $C_N=\frac{14}{3}$ and $C_\Delta=\frac{4}{3}$ are obtained by taking the expectation value of the spin-flavor operator $(\vec{\sigma}_i\cdot\vec{\sigma}_j)(\vec{\lambda}_i^F\cdot\vec{\lambda}_j^F)$ in the appropriate wave functions. V^0 includes both the linear confining potential and the spin-independent part of the q-q potential

$$V_{qq}(r_{ij})=V^0(r_{ij})+V^S(r_{ij})\vec{\sigma}_i.\vec{\sigma}_j+V^X(r_{ij})(\vec{\sigma}_i.\vec{\sigma}_j)(\vec{\lambda}_i^F.\vec{\lambda}_j^F). \tag{4.41}$$

Equation (4.39) is solved numerically, subject to appropriate boundary conditions, for masses and wave functions of ground and first radial excitations of the S, P, and D waves of the nucleon and Δ. Numerical results and discussions can be found in Refs. [14, 15].

The hyperspherical harmonics technique has also been used in the study of identical flavor four-quark systems [20]. Recently, the three-body wave equation with the constituent quarks bound by a suitable hypercentral potential has been solved by Abou-Salem to obtain resonance states of N, Δ, Λ, and Σ baryon systems [21].

4.6.2 Nuclear Few-Body Systems

Several nuclear few-body systems are of great interest in physics. The trinucleon systems (^{3}H and [3]He nuclei) are essential for the investigation of nuclear three-body force. We will discuss them in detail in Chap. 5. Some heavier nuclei can be treated as few-body nuclear systems with clusters as building blocks. Important among these are the halo nuclei containing a stable core and a few valence nucleons. As an example [11]Li nucleus can be treated as [9]Li core and two loosely bound valence neutrons [22]. In halo nuclei, very weakly bound extra-core nucleons form a low-density halo around the stable core giving the nucleus a larger radius than neighboring stable nuclei. The alpha (α) particle is the [4]He nucleus and is a four-body system, which will be discussed below. The α particle is very stable and it has a large binding energy. Hence $A=4n$ (with $n=3,4,\ldots$) nuclei can be treated as clusters of α particles. Such clusters have also been studied as few-body systems using HHEM. This technique can also be applied in the analysis of nuclear scattering [23]. $N-d$ scattering has been treated as a three-body system, where the deuteron is treated as a bound state of a proton and a neutron [24, 25].

Halo Nuclei

Experimental discovery of halo nuclei became possible in the 1980s with the development of intense radioactive ion beam facility. These nuclei have one or more nucleons very weakly bound to a fairly stable core. In such nuclei none of the two-body subsystems is bound. Importance of halo nuclei lies in the investigation of longer range nuclear interaction in low-density regime. The core can be treated as a particle, hence a few-body model is possible. With large neutron excess, such light nuclei lie near the neutron drip line, giving a small separation energy for the valence nucleons. The first experiments in the mid-1980s observed large interaction or reaction cross sections at high or low energies for light neutron drip line nuclei ^{11}Li, ^{11}Be, ^{14}Be, and ^{17}B. This gives an abnormally large cross section for the halo nucleus. The valence nucleons are thus much further from the core than the average separation in a stable nucleus. Hence they form a very low-density *halo* around the core, giving the halo nucleus a larger r.m.s. radius than the $r_0 A^{\frac{1}{3}}$ formula. A very narrow ^{9}Li transverse momentum distribution was found from ^{11}Li fragmentation at high energy, giving evidence of a halo structure and justification of a few-body model consisting of the core and valence neutrons [26–28]. The halo nuclei ^{6}He and ^{6}Li can be treated as $(\alpha + N + N)$ and $(\alpha + P + N)$, respectively. However in the latter the finite tail of the deuteron wave function gives a finite probability of the α plus deuteron channel [26]. Bound states of two neutron halo nuclei ^{6}He and ^{11}Li were investigated successfully by HHEM [26, 29, 30]. Binding energy and other observables calculated by the HHEM agree well with experimental results and other calculations.

Resonance States of Halo Nuclei

Resonance states of the halo nuclei have also been studied using the hyperspherical formalism. However, such resonances are produced by extremely shallow effective potentials. Hence they are very broad and lie well into the continuum. Consequently, the wave function is not localized and standard numerical solution of the Schrödinger equation is not sufficiently accurate. Using a novel technique of the supersymmetric quantum mechanics the shallow potential is replaced by an isospectral potential, which can be made desirably narrow and deep, leading to desired precision [31]. This method needs the existence of a ground state with the same quantum numbers as the resonance investigated. But this method cannot be used if there is no lower energy bound state with the same quantum numbers as the resonance studied. In such a case, one can construct a *bound state in the continuum* technique to produce the isospectral potential [31]. The supersymmetric isospectral potential constructed using the hyperspherical technique has been used in the study of resonances in $A = 6$ nuclei [32]. For the 2^+ resonance of ^{6}He and the $\frac{5}{2}^+$ resonance state of ^{11}Be the method of bound state in continuum using HHEM has been used [33, 34], as there is no corresponding lower energy bound state. In all these cases excellent agreements with experimental results have been obtained.

Bound State of Four Nucleon: The α Particle

The alpha (α) particle is the bound ground state of the four nucleon system, consisting of two protons and two neutrons, i.e., the ^{4}He nucleus. Its binding energy is about 28.4 MeV. The binding energy per particle is much higher than that of the neighboring nuclei. Hence it occupies a special place, since for low energy nuclear physics, it can be considered as a 'particle'. In the shell model description, it is the first doubly magic nucleus, fully occupying the $1s\frac{1}{2}$ shell for both proton and neutron. This explains the large binding energy. Although the net Coulomb force (only between protons) is repulsive, the strong nucleon–nucleon force, which is dominantly attractive, produces the strong binding. As we have discussed earlier, a full quantum mechanical treatment of the four-body system is a lot more complicated than the three-body system. The α particle being a doubly magic $1s$-shell nucleus, its ground state has $J^\pi = 0^+$, with both total spin and total orbital angular momenta taking zero values ($s = 0, l = 0$) for the ground state in the shell model description. Hence it behaves as a spin zero boson. Since both proton and neutron are spin-half fermions, the total wave function of the α particle is antisymmetric under any pair exchange (including spin and isospin degrees). Clearly, the ground state can have contributions from various mixed symmetry spin-isospin states, combined with space wave function of conjugate symmetry, such that the combination of space-spin-isospin wave functions is totally antisymmetric. From the shell model description, the ground state of ^{4}He nucleus is the fully antisymmetric determinantal wave function constructed out of single particle wave functions $\phi_{nljm_jm_t}(i)$ with $i = 1, 4$ for four nucleons (m_t is the projection of isospin with the values $\frac{1}{2}$ and $-\frac{1}{2}$ for proton and neutron, respectively). This state has $s = 0, l = 0$, with the space part of the wave function being totally symmetric under any pair exchange. Since the shell model is a very good starting point, the contribution of the space totally symmetric S ($l = 0$) state is very large and to a very good approximation, the α particle can be represented by this state alone.

Thus in the HHEM, one can restrict the symmetry components to the space totally symmetric wave function only. The space wave function is expanded in the totally symmetric HH basis for the four-body system. The latter is obtained by the method of Sect. 4.3.1 [35, 36]. For the ground state the L_m approximation was found to be adequate. But for the 0^+ excited state it was necessary to include higher L values for a proper convergence. Charge form factors for e scattering from ^{4}He was calculated by Sanyal and Mukherjee [36] using the HHEM technique.

Exotic Few-Body Nuclear Systems

Exotic nuclei containing a few hyperons have been modeled as a system consisting of a normal nucleus as a core and a few hyperons. For example, the exotic nucleus $^6_{\Lambda\Lambda}$He is considered as a bound state of the alpha particle and two Λ particles. Its ground state has been investigated by hyperspherical technique [37]. Ground state structure, $\Lambda\Lambda$ dynamics, and hyperon–nucleon interaction were also studied for low and medium mass hypernuclei [38]. Techniques of supersymmetric quantum mechanics were used in conjunction with HHEM to study excited states of such exotic nuclei [39].

Nuclear and Atomic Clusters

Some nuclei have been treated as a bound state of subsystems to gain a specific insight. For example, the ^{5}H nucleus (with large neutron excess) has been considered as a three-body system of ^{3}H and two neutrons [40]. The continuum resonance spectrum of this nucleus has been investigated using the complex-scaled hyperspherical adiabatic expansion (see Chap. 10) method [41].

The alpha particle model has been proposed for N = Z nuclei with A = 4n, for $n = 3, 4, \ldots$. It is another example of nuclear cluster. In this model nuclei like ^{12}C, ^{16}O, etc. are considered as bound states of three, four, etc. α particles. This is justified because the α particle has a large binding energy, compared to the binding energy of neighboring nuclei. Such nuclei have been treated by the HHEM [42, 43].

Properties of Helium and Rubidium atomic clusters (trimer, tetramer, etc.) have also been calculated using the HHEM [44], hyperspherical Monte Carlo description [45], adiabatic hyperspherical method [46], and by variational calculation using correlated HH basis [47]. The effective He–He interaction was chosen to be the standard TTY potential. The hyperspherical technique has also been used to calculate binding energy and scattering observables of ^{4}He trimers [48, 49].

Supersymmetric isospectral formalism [31] together with the HHEM was used for near-zero energy states of He clusters [50]. Such states are spatially very extended and hence the convergence of the HH expansion is very slow. As a result accurate calculation of the wave function in the asymptotic region is very difficult. This difficulty is avoided by constructing the isospectral potential according to the prescription of supersymmetric quantum mechanics (SSQM). This isospectral potential has the property that its spectrum is identical with the original spectrum, but the width of the isospectral potential can be controlled through a suitable but arbitrary parameter [31]. This parameter is chosen such that the isospectral potential has a narrow and deep well. Hence states in this well are not too extended and a high precision in the observables is possible. The hyperspherical technique has also been used in the study of diffuse Rubidium clusters, where Rb–Rb interaction is given by the s-wave scattering length (a_s) of Rb atoms [51]. The level spacing statistics and spectral correlations of diffuse van der Waals clusters are investigated using the HHEM [52].

References

1. Krivec, R.: Few-Body Syst. **25**, 199 (1998)
2. Avery, J.: J. Phys. Chem. **97**, 2406 (1993); Avery, J.: J. Comp. Appl. Math. **233**, 1366 (2010)
3. Meremianin, A.V.: J. Math. Phys. **50**, 013526 (2009)
4. Wang, D., Kuppermann, A.: J. Phys. Chem. A **113**, 15384 (2009); Kuppermann, A.: Phys. Chem. Chem. Phys. **13**, 8259 (2011)
5. Zernike, F., Brinkman, H.C.: Proc. Kon. Ned. Acad. Wet. **33**, 3 (1935)
6. Ballot, J.L., Fabre de la Ripelle, M.: Ann. Phys. (N.Y.) **127**, 62 (1980)
7. Abramowitz, M., Stegun, I.A.: Handbook of Mathematical Functions. National Institute of Standards and Technology, USA (1964)

8. Gattobigio, M., Kievsky, A., Viviani, M., Barletta, P.: Phys. Rev. A **79**, 032513 (2009); Gattobigio, M., Kievsky, A., Viviani, M.: Phys. Rev. C **83**, 024001 (2011); Deflorian, S., et al, EPJ Web Conf. **66**, 02025 (2014)
9. Fabre de la Ripelle, M.: Ann. Phys. (N.Y.) **147**, 281 (1983)
10. Rose, M.E.: Elementary Theory of Angular Momentum. Wiley, New York (1967)
11. Edmonds, A.R.: Angular Momentum in Quantum Mechanics. Princeton University Press, Princeton (1957)
12. Chattopadhyay, R., Das, T.K.: Fiz. A **5**, 11 (1996)
13. Richard, J.-M.: Phys. Rep. **212**, 1 (1992); Vassallo, A.: Baryon structure with hypercentral constituent quark model. Ph.D. Thesis, Universita degli Studi di Genova (2004)
14. Dziembowski, Z., Fabre de la Ripelle, M., Miller, G.A.: Phys. Rev. C **53**, R2038 (1996)
15. Nag, R., Sanyal, S., Mukherjee, S.N.: Phys. Rev. D **36**, 2788 (1987); Nag, R., Sanyal, S., Mukherjee, S.N.: Prog. Theor. Phys. **83**, 51 (1990); Nag, R., Mukherjee, S.N., Sanyal, S.: Phys. Rev. C **44**, 1709 (1991); Nag, R., Sanyal, S.: J. Phys. G **20**, L125 (1994); Sanyal, S., Nag, R.: J. Phys. G **22**, L7 (1996); Sanyal, S., Nag, R.: Phys. Rev. D **59**, 014027 (1998)
16. Garcilazo, H., Vijande, J., Valcarce, A.: J. Phys. G: Nucl. Part. Phys. **34**, 961 (2007)
17. Boroun, G.R., Abdolmaleki, H.: Phys. Scr. **80**, 065003 (2009)
18. Verma, S.P., Lall, N.R.: Indian J. Pure Appl. Phys. **48**, 851 (2010)
19. Rezaei, B., Boroun, G.R., Abdolmaleki, H.: Int. J. Math. Phys. Quan. Eng. **7**, 446 (2013)
20. Vijande, J., Barnea, N., Valearce, A.: Nucl. Phys. A **790**, 542c (2007)
21. Abou-Salem, L.I.: Adv. High Energy Phys. **2014**, Article ID 196434 (2014)
22. Al-Khalili, J.: An Introduction to Halo Nuclei, in Al-Khalili, J., Roecki, E. (eds.) The Euroschool Lectures on Physics with Exotic Beams, vol. I, Lecture Notes in Physics 651. Springer, Berlin, 2004
23. Danilin, B.V., et al.: Phys. Rev. C **55**, R577 (1997)
24. Marcucci, L.E., et al.: Phys. Rev. C **80**, 034003 (2009)
25. Kievsky, A., et al.: Phys. Rev. C **81**, 044003 (2010); J. Phys. G **35**, 063101 (2008)
26. Zhukov, M.V., et al.: Phys. Rep. **231**, 151 (1993)
27. Ershov, S.N., Vaagen, J.S., Zhukov, M.V.: Phys. Rev. C **86**, 034331 (2012)
28. Shulgina, N.B., Jonson, B., Zhukov, M.V.: Nucl. Phys. A **825**, 175 (2009)
29. Khan, M.A., Dutta, S.K., Das, T.K., Pal, M.K.: J. Phys. G **24**, 1519 (1998); Abdul Khan, Md., Das, T.K.: Pramana (J. Phys.) **57**, 701 (2001)
30. Bacca, S., Barnea, N., Schwenk, A.: Phys. Rev. C **86**, 034321 (2012)
31. Cooper, F., Khare, A., Sukhatme, U.: Phys. Rep. **251**, 267 (1995)
32. Dutta, S.K., Das, T.K., Khan, M.A., Chakrabarti, B.: Few-Body Syst. **35**, 33 (2004); Dutta, S.K., Das, T.K., Khan, M.A., Chakrabarti, B.: Int. J. Mod. Phys. E **13**, 811 (2004)
33. Dutta, S.K., Das, T.K., Khan, M.A., Chakrabarti, B.: J. Phys. G: Nucl. Part. **29**, 2411 (2003)
34. Mahapatra, S., Das, T.K., Dutta, S.K.: Int. J. Mod. Phys. E **18**, 1741 (2009)
35. Elminyawi, I., Levinger, J.S.: Phys. Rev. C **28**, 82 (1983)
36. Sanyal, S., Mukherjee, S.N.: Phys. Rev. C **29**, 665 (1984); ibid **31**, 33 (1985); ibid **36**, 67 (1987)
37. Khan, Md.A, Das, T.K.: Fiz. B **9**, 55 (2000)
38. Khan, Md.A., Das, T.K.: Pramana (J. Phys.) **56**, 57 (2001); Abdul Khan, Md., Das, T.K.: Fiz. B **10**, 83 (2001)
39. Khan, MdA, Das, T.K., Chakrabarti, B.: Int. J. Mod. Phys. E **10**, 107 (2001)
40. Shulgina, N.B., Danilin, B.V., Grigorenko, L.V., Zhukov, M.V., Bang, J.M.: Phys. Rev. C **62**, 014312 (2000)
41. de Diego, R., Garrido, E., Fedorov, D.V., Jensen, A.S.: Nucl. Phys. A **786**, 71 (2007)
42. Vallieres, M., Coelho, H.T., Das, T.K.: Nucl. Phys. A **271**, 95 (1976); Consoni, L., Das, T.K., Coelho, H.T.: Rev. Bras. Fis. **13**, 137 (1983)
43. Nguyen, N.B., Nunes, F.M., Thompson, I.J., Brown, E.F.: Phys. Rev. Lett. **109**, 141101 (2012)
44. Das, T.K., Chakrabarti, B., Canuto, S.: J. Chem. Phys. **134**, 164106 (2011)
45. Blume, D., Greene, C.H.: J. Chem. Phys. **112**, 8053 (2000)
46. Suno, H., Esry, B.D.: Phys. Rev. A **78**, 062701 (2008)
47. Barletta, P., Kievsky, A.: Phys. Rev. A **64**, 042514 (2001)

48. Motovilov, A.K., et al.: Eur. Phys. J. D **13**, 33 (2001)
49. Sandhas, W., et al.: Few-Body Syst. **34**, 137 (2004)
50. Haldar, S.K., Chakrabarti, B., Das, T.K.: Few-Body Syst. **53**, 283 (2012)
51. Debnath, P.K., Chakrabarti, B., Das, T.K., Canuto, S.: J. Chem. Phys **137**, 014301 (2012)
52. Haldar, S.K., Chakrabarti, B., Chavda, N.D., Das, T.K., Canuto, S., Kota, V.K.B.: Phys. Rev. A **89**, 043607 (2014)

Chapter 5
The Trinucleon System

Abstract Solutions of trinucleon nuclei (^{3}H and ^{3}He) are important in the study of nuclear forces. The ground-state wave function, including spin and isospin, is antisymmetrized, giving rise to several components. Next the optimal subset for the trinucleon interacting via central and tensor interactions is constructed. The potential matrix elements are calculated using geometrical structure coefficients. The latter coefficients (for coupling of various components of the wave function through central and tensor forces) are obtained elegantly by solving a set of linear inhomogeneous equations. Results of typical calculations are presented as illustration. Effect of nuclear three-body force is also discussed.

In this chapter, we discuss the trinucleon system in detail. There are two bound trinucleon nuclei: ^{3}H containing a proton and two neutrons, and ^{3}He containing two protons and a neutron, having binding energies of 8.48 MeV and 7.73 Mev, respectively. These are the simplest nuclei after the deuteron ^{2}H. The trinucleon nuclei have special importance in nuclear physics, since these can provide information on the existence and importance of nuclear three-body force. Being three-body systems, essentially exact calculations are possible. In Chap. 3, we discussed how the hyperspherical harmonics expansion method can be developed for the three-body system. In this chapter, we will discuss how appropriate symmetry can be imposed and obtain the resultant coupled differential equations.

A trinucleon nucleus is a bound state of three spin $\frac{1}{2}$ fermions. In the simplest shell model picture, two identical nucleons (two protons for ^{3}He and two neutrons for ^{3}H) occupy a closed $1s\frac{1}{2}$ shell and the third nucleon occupies $1s\frac{1}{2}$ shell of the other nucleon type. Hence, the spin parity is $J^\pi = \frac{1}{2}^+$, which agrees with experiment. In this picture, total spin and orbital angular momenta are good and have values $s = \frac{1}{2}$ and $l = 0$. However, from two-nucleon studies, it is known that nuclear force has a small noncentral contribution and has exchange character. Hence, l and s are not strictly good quantum numbers, although the $l = 0, s = \frac{1}{2}$ (called the S state of the trinucleon) has a dominant contribution. Introduction of the isospin makes the proton and the neutron identical spin $\frac{1}{2}$ fermions. Hence, the full trinucleon wave function must be antisymmetric under exchange of any pair. The S state corresponds to the

© Springer India 2016

T.K. Das, *Hyperspherical Harmonics Expansion Techniques,*

Theoretical and Mathematical Physics, DOI 10.1007/978-81-322-2361-0_5

space totally symmetric state combined with the totally antisymmetric spin–isospin wave function. For the three-body system, it is possible to have space and spin–isospin states of mixed conjugate symmetry, so that the full wave function is totally antisymmetric. In the following, we discuss how these states can be constructed. In Chap. 3, Sect. 3.5, we discussed the symmetrization of the spatial wave function. Following the work of Ballot and Fabre [1], we discuss the symmetrization of the spin–isospin wave function of the trinucleon in Sect. 5.1. In this chapter, we will assume the potential to have a soft core, as is usually chosen for numerical calculation. The three-body hyperspherical harmonics technique has to be modified for potentials with a hard core [2].

5.1   Symmetrization of Spin–Isospin Wave Function

The total spin (s) of three spin $\frac{1}{2}$ nucleons can be $\frac{3}{2}$ (once), $\frac{1}{2}$ (twice). Likewise, the total isospin (t) of three isospin $\frac{1}{2}$ particles can be $\frac{3}{2}$ (once), $\frac{1}{2}$ (twice). The normalized spin wave function of particles i, j and k is

$$
|(s_{ij}\frac{1}{2})sm_s\rangle_k^{(S)} = \sum_{\nu_1\nu_2\nu_3} (-1)^{\frac{1}{2}-s_{ij}-\nu-m_s} \hat{s}\hat{s}_{ij} \begin{pmatrix} \frac{1}{2} & \frac{1}{2} & s_{ij} \\ \nu_1 & \nu_2 & -\nu \end{pmatrix} \begin{pmatrix} s_{ij} & \frac{1}{2} & s \\ \nu & \nu_3 & -m_s \end{pmatrix}
$$
$$
\chi_{\nu_1}^{\frac{1}{2}}(i)\,\chi_{\nu_2}^{\frac{1}{2}}(j)\,\chi_{\nu_3}^{\frac{1}{2}}(k), \tag{5.1}
$$

where $\vec{s}_{ij} = \vec{s}_i + \vec{s}_j$ and the total spin is $\vec{s} = \vec{s}_{ij} + \vec{s}_k$. In Eq. (5.1), $\hat{s}$ stands for $(2s+1)^{\frac{1}{2}}$ and $\chi_\nu^{\frac{1}{2}}(i)$ is the spin wave function of the i-th particle with projection ν. The matrix notation is a 3-j symbol. The subscript of the ket vector on left side represents the last coupled particle (k) of the triplet (i, j, k) and a superscript (S) indicates spin wave function. This wave function is symmetric or antisymmetric under $i \leftrightarrow j$ for $s_{ij} = 1$ or 0 and is denoted by the abbreviated notation $|+\rangle_k^{(S)}$ and $|-\rangle_k^{(S)}$, respectively. A similar expression gives the total isospin wave function of three isospin $\frac{1}{2}$ particles:

$$
|(t_{ij}\frac{1}{2})tm_t\rangle_k^{(T)} = \sum_{\mu_1\mu_2\mu_3} (-1)^{\frac{1}{2}-t_{ij}-\mu-m_t} \hat{t}\hat{t}_{ij} \begin{pmatrix} \frac{1}{2} & \frac{1}{2} & t_{ij} \\ \mu_1 & \mu_2 & -\mu \end{pmatrix} \begin{pmatrix} t_{ij} & \frac{1}{2} & t \\ \mu & \mu_3 & -m_t \end{pmatrix}
$$
$$
\tau_{\mu_1}^{\frac{1}{2}}(i)\,\tau_{\mu_2}^{\frac{1}{2}}(j)\,\tau_{\mu_3}^{\frac{1}{2}}(k), \tag{5.2}
$$

where isospins of particles i and j are coupled to an intermediate isospin $\vec{t}_{ij}$, which in turn is coupled to $\vec{t}_k$ to a resultant isospin of $\vec{t}$ and $\tau_\mu^{\frac{1}{2}}(i)$ is the isospin wave function of the i particle. A superscript (T) indicates isospin wave function.

The known quantum numbers of the ground state of trinucleon are $J^\pi = \frac{1}{2}^+$. Three spin halves can be combined as $\vec{s}_i + \vec{s}_j = \vec{s}_{ij}$, $\vec{s}_{ij} + \vec{s}_k = \vec{s}$ with $s_{ij} = 0$ or 1 and $s = \frac{1}{2}, \frac{1}{2}$, and $\frac{3}{2}$. The spin state with $s = \frac{3}{2}$ is totally symmetric under any pair exchange and the two $s = \frac{1}{2}$ spin states correspond to mixed symmetry spin states under the exchange of a given pair. An identical result is valid for the isospin under exchanges in isospin space. One can combine states in isospin and spin spaces to get one totally antisymmetric, one totally symmetric, and two mixed symmetry states under combined exchanges in isospin–spin spaces. These states are to be combined with space wave functions of conjugate symmetry, so that the combination has the desired symmetry, *viz.*, totally antisymmetric under any pair exchange in the combined space–isospin–spin spaces.

Since $J = \frac{1}{2}$ and s can have values $\frac{1}{2}$ (twice) and $\frac{3}{2}$ (once), the possible values of total orbital angular momentum (l) of the system are 0, 1, and 2. Information from experiments shows that the ground state has a dominant contribution from the space totally symmetric $l = 0$ state (S) and smaller contributions from space mixed symmetry $l = 0$ state (S') and a $l = 2$ state (D). In the following subsections, we obtain different possible isospin–spin states of the trinucleon.

5.1.1 States Having Total Isospin $t = \frac{1}{2}$ and Spin $s = \frac{1}{2}$

For $t = \frac{1}{2}, s = \frac{1}{2}$ states, we introduce a parametric angle ϖ (as was done for the space wave function in Chap. 3, Sect. 3.5) and make a l.c. of the symmetric and antisymmetric states under $i \leftrightarrow j$

$$|W(\varpi)\rangle_k^{(X)} = \sin \varpi \left| \left(0\frac{1}{2}\right) \frac{1}{2} m \right\rangle_k^{(X)} + \cos \varpi \left| \left(1\frac{1}{2}\right) \frac{1}{2} m \right\rangle_k^{(X)}, \qquad (5.3)$$

where the superscript (X) stands for (S) or (T) for spin or isospin, respectively. Note that $|W(\varpi)\rangle_k^{(X)}$ is normalized. We can easily verify that following l.c. are antisymmetric under the exchange of angular momenta of indicated particles

$$
\begin{aligned}
|W(\pi/2)\rangle_k^{(X)} &= \left| \left(0\frac{1}{2}\right) \frac{1}{2} m \right\rangle_k^{(X)} \equiv |-\rangle_k^{(X)} & \text{antisymm for } i \leftrightarrow j \\
|W(\pi/2 - 2\pi/3)\rangle_k^{(X)} &= \frac{\sqrt{3}}{2} |+\rangle_k^{(X)} - \frac{1}{2} |-\rangle_k^{(X)} & \\
&= \left| \left(0\frac{1}{2}\right) \frac{1}{2} m \right\rangle_i^{(X)} & \text{antisymm for } j \leftrightarrow k \quad (5.4) \\
|W(\pi/2 + 2\pi/3)\rangle_k^{(X)} &= -\frac{\sqrt{3}}{2} |+\rangle_k^{(X)} - \frac{1}{2} |-\rangle_k^{(X)} & \\
&= \left| \left(0\frac{1}{2}\right) \frac{1}{2} m \right\rangle_j^{(X)} & \text{antisymm for } k \leftrightarrow i
\end{aligned}
$$

In going from first to second line of the last two equations of Eq. (5.4), properties under exchange of individual angular moments of the coupled state of three angular momenta $|(j_1 j_2) j_{12} j_3; jm\rangle$ have been used [3]. In a similar fashion, we have the following l.c. which are symmetric under exchange of the first two angular momenta of the indicated particles

$$|W(0)\rangle_k^{(X)} \quad = |(1\tfrac{1}{2})\tfrac{1}{2}m\rangle_k^{(X)} \equiv |+\rangle_k^{(X)} \qquad \text{symm for } i \leftrightarrow j$$

$$|W(-2\pi/3)\rangle_k^{(X)} = -\tfrac{1}{2}|+\rangle_k^{(X)} - \tfrac{\sqrt{3}}{2}|-\rangle_k^{(X)}$$

$$= |(1\tfrac{1}{2})\tfrac{1}{2}m\rangle_i^{(X)} \qquad \text{symm for } j \leftrightarrow k \qquad (5.5)$$

$$|W(2\pi/3)\rangle_k^{(X)} \quad = -\tfrac{1}{2}|+\rangle_k^{(X)} + \tfrac{\sqrt{3}}{2}|-\rangle_k^{(X)}$$

$$= |(1\tfrac{1}{2})\tfrac{1}{2}m\rangle_j^{(X)} \qquad \text{symm for } k \leftrightarrow i$$

The combinations $|W(\varpi)\rangle_k^{(T)} \, |W(\varpi - \pi/2)\rangle_k^{(S)}$ and $|W(\varpi - \pi/2)\rangle_k^{(T)} \, |W(\varpi)\rangle_k^{(S)}$ are both antisymmetric in spin–isospin space under $i \leftrightarrow j$, $j \leftrightarrow k$ and $k \leftrightarrow i$ for $\varpi = \pi/2$, $\pi/2 - 2\pi/3$ and $\pi/2 + 2\pi/3$, respectively. Using Eq. (5.3), we have

$$|W(\varpi)\rangle_k^{(T)} \, |W(\varpi - \pi/2)\rangle_k^{(S)} = -\sin\varpi \cos\varpi \left[|-\rangle_k^{(T)}|-\rangle_k^{(S)} - |+\rangle_k^{(T)}|+\rangle_k^{(S)}\right]$$

$$+ \sin^2\varpi |-\rangle_k^{(T)}|+\rangle_k^{(S)} - \cos^2\varpi|+\rangle_k^{(T)}|-\rangle_k^{(S)} \quad (5.6)$$

and

$$|W(\varpi - \pi/2)\rangle_k^{(T)} \, |W(\varpi)\rangle_k^{(S)} = -\sin\varpi \cos\varpi \left[|-\rangle_k^{(T)}|-\rangle_k^{(S)} - |+\rangle_k^{(T)}|+\rangle_k^{(S)}\right]$$

$$+ \sin^2\varpi |+\rangle_k^{(T)}|-\rangle_k^{(S)} - \cos^2\varpi|-\rangle_k^{(T)}|+\rangle_k^{(S)} \quad (5.7)$$

In a similar fashion, we can construct two combinations of isospin and spin wave functions, which are symmetric under exchange $i \leftrightarrow j$, $j \leftrightarrow k$ and $k \leftrightarrow i$ for $\varpi = \pi/2$, $\pi/2 - 2\pi/3$ and $\pi/2 + 2\pi/3$, respectively,

$$|W(\varpi)\rangle_k^{(T)} \, |W(\varpi)\rangle_k^{(S)} = \sin\varpi \cos\varpi \left[|+\rangle_k^{(T)}|-\rangle_k^{(S)} + |-\rangle_k^{(T)}|+\rangle_k^{(S)}\right]$$

$$+ \sin^2\varpi |-\rangle_k^{(T)}|-\rangle_k^{(S)} + \cos^2\varpi|+\rangle_k^{(T)}|+\rangle_k^{(S)} \quad (5.8)$$

and

$$|W(\varpi - \pi/2)\rangle_k^{(T)} \, |W(\varpi - \pi/2)\rangle_k^{(S)} = -\sin\varpi \cos\varpi \left[|+\rangle_k^{(T)}|-\rangle_k^{(S)} + |-\rangle_k^{(T)}|+\rangle_k^{(S)}\right]$$

$$+ \sin^2\varpi |+\rangle_k^{(T)}|+\rangle_k^{(S)} + \cos^2\varpi|-\rangle_k^{(T)}|-\rangle_k^{(S)}. \quad (5.9)$$

Equations (5.6)–(5.9) give combinations of isospin–spin wave functions which have specified symmetry under a pair exchange according to the specific values of ϖ. If we can find a l.c. of these which is independent of ϖ, then it will have a specified symmetry under exchange of any pair of particles. We can find two such normalized combinations from Eqs. (5.6) and (5.7)

$$\frac{1}{\sqrt{2}}\left[-|W(\varpi)\rangle_k^{(T)} \, |W(\varpi - \pi/2)\rangle_k^{(S)} + |W(\varpi - \pi/2)\rangle_k^{(T)} \, |W(\varpi)\rangle_k^{(S)}\right]$$

$$= \frac{1}{\sqrt{2}}\left[|+\rangle_k^{(T)}|-\rangle_k^{(S)} - |-\rangle_k^{(T)}|+\rangle_k^{(S)}\right] \equiv \Gamma_{\frac{1}{2}\frac{1}{2}}(A), \quad (5.10)$$

which is totally antisymmetric in isospin–spin space under any pair exchange. In Eq. (5.10), we introduce the notation $\Gamma_{ts}(\mathcal{R})$ as the isospin–spin state of total isospin t and total spin s, having a particular symmetry $\mathcal{R}$ under pair exchanges. In a similar manner, we have from Eqs. (5.8) and (5.9)

$$\frac{1}{\sqrt{2}}\left[|W(\varpi)\rangle_k^{(T)}\,|W(\varpi)\rangle_k^{(S)} + |W(\varpi - \pi/2)\rangle_k^{(T)}\,|W(\varpi - \pi/2)\rangle_k^{(S)}\right]$$

$$= \frac{1}{\sqrt{2}}\left[|+\rangle_k^{(T)}|+\rangle_k^{(S)} + |-\rangle_k^{(T)}|-\rangle_k^{(S)}\right] \equiv \Gamma_{\frac{1}{2}\frac{1}{2}}(S), \tag{5.11}$$

which is totally symmetric in isospin–spin space under any pair exchange.

Two more mixed symmetry (hence dependent on ϖ) combinations can be constructed. From Eqs. (5.6) and (5.7), we have

$$\frac{1}{\sqrt{2}}\left[|W(\varpi)\rangle_k^{(T)}\,|W(\varpi - \pi/2)\rangle_k^{(S)} + |W(\varpi - \pi/2)\rangle_k^{(T)}\,|W(\varpi)\rangle_k^{(S)}\right]$$

$$= \sin 2\varpi\,\frac{1}{\sqrt{2}}\left[|+\rangle_k^{(T)}|+\rangle_k^{(S)} - |-\rangle_k^{(T)}|-\rangle_k^{(S)}\right]$$

$$- \cos 2\varpi\,\frac{1}{\sqrt{2}}\left[|+\rangle_k^{(T)}|-\rangle_k^{(S)} + |-\rangle_k^{(T)}|+\rangle_k^{(S)}\right]$$

$$\equiv \sin 2\varpi\,\Gamma_{\frac{1}{2}\frac{1}{2}}(M+) - \cos 2\varpi\,\Gamma_{\frac{1}{2}\frac{1}{2}}(M-). \tag{5.12}$$

This combination is antisymmetric in isospin–spin space under exchange of particles $i \leftrightarrow j$, $j \leftrightarrow k$ and $k \leftrightarrow i$, respectively, for $\varpi = \pi/2$, $\pi/2 - 2\pi/3$ and $\pi/2 + 2\pi/3$. The arguments $(M+)$ and $(M-)$ of $\Gamma_{\frac{1}{2}\frac{1}{2}}$ refer to mixed symmetry for $t = \frac{1}{2}$, $s = \frac{1}{2}$.

We have from Eqs. (5.8) and (5.9) that

$$\frac{1}{\sqrt{2}}\left[-|W(\varpi)\rangle_k^{(T)}\,|W(\varpi)\rangle_k^{(S)} + |W(\varpi - \pi/2)\rangle_k^{(T)}\,|W(\varpi - \pi/2)\rangle_k^{(S)}\right]$$

$$= -\sin 2\varpi\,\frac{1}{\sqrt{2}}\left[|+\rangle_k^{(T)}|-\rangle_k^{(S)} + |-\rangle_k^{(T)}|+\rangle_k^{(S)}\right]$$

$$- \cos 2\varpi\,\frac{1}{\sqrt{2}}\left[|+\rangle_k^{(T)}|+\rangle_k^{(S)} - |-\rangle_k^{(T)}|-\rangle_k^{(S)}\right]$$

$$\equiv -\sin 2\varpi\,\Gamma_{\frac{1}{2}\frac{1}{2}}(M-) - \cos 2\varpi\,\Gamma_{\frac{1}{2}\frac{1}{2}}(M+), \tag{5.13}$$

which is symmetric in isospin–spin space under exchange of particles $i \leftrightarrow j$, $j \leftrightarrow k$, and $k \leftrightarrow i$, respectively, for $\varpi = \pi/2$, $\pi/2 - 2\pi/3$, and $\pi/2 + 2\pi/3$.

Thus for $t = \frac{1}{2}$ and $s = \frac{1}{2}$ there are four possibilities of isospin–spin states $(\Gamma_{ts}(\mathcal{R})$, where $\mathcal{R}$ denotes a symmetry), viz., one totally antisymmetric $(\Gamma_{\frac{1}{2}\frac{1}{2}}(A))$, one totally symmetric $(\Gamma_{\frac{1}{2}\frac{1}{2}}(S))$, and two mixed symmetry $(\Gamma_{\frac{1}{2}\frac{1}{2}}$ $(M+)$ and $\Gamma_{\frac{1}{2}\frac{1}{2}}$ $(M-))$ states. These are to be combined with space wave function of conjugate symmetry to obtain the desired symmetry of the total wave function, which we will discuss in Sect. 5.2.

5.1.2   States Having Total Isospin $t = \frac{3}{2}$ and Spin $s = \frac{1}{2}$

The isospin wave function with $t = \frac{3}{2}$ is totally symmetric under any pair exchange in isospin space. We denote this state by (see Eq. 5.2)

$$\left| \left(1 \frac{1}{2} \right) \frac{3}{2} m_t \right\rangle_k^{(T)} \equiv \left| 0 \right\rangle^{(T)}. \tag{5.14}$$

The subscript is omitted since it is the same for all k. Hence, the symmetry property under exchange of particles in the isospin–spin space is given by that in spin space only, which are given by Eqs. (5.4) and (5.5) for $(X) = (S)$. For the same set of values of ϖ introduced earlier, we have (using our previous notation for spin states) from Eq. (5.3)

$$|0\rangle^{(T)} |W(\varpi)\rangle_k^{(S)} = \sin \varpi \, |0\rangle^{(T)} |-\rangle_k^{(S)} + \cos \varpi |0\rangle^{(T)} |+\rangle_k^{(S)}$$
$$\equiv \sin \varpi \, \Gamma_{\frac{3}{2}\frac{1}{2}}(M'-) + \cos \varpi \, \Gamma_{\frac{3}{2}\frac{1}{2}}(M'+), \tag{5.15}$$

which is antisymmetric under exchanges $i \leftrightarrow j$, $j \leftrightarrow k$ and $k \leftrightarrow i$ for $\varpi = \pi/2$, $\pi/2 - 3\pi/2$ and $\pi/2 + 3\pi/2$, respectively. The argument $(M'\pm)$ of $\Gamma_{\frac{3}{2}\frac{1}{2}}$ refers to mixed symmetry for $t = \frac{3}{2}, s = \frac{1}{2}$. Similarly, we also have

$$|0\rangle^{(T)} |W(\varpi - \pi/2)\rangle_k^{(S)} = \sin \varpi \, |0\rangle^{(T)} |+\rangle_k^{(S)} - \cos \varpi |0\rangle^{(T)} |-\rangle_k^{(S)}$$
$$\equiv - \cos \varpi \, \Gamma_{\frac{3}{2}\frac{1}{2}}(M'-) + \sin \varpi \, \Gamma_{\frac{3}{2}\frac{1}{2}}(M'+), \tag{5.16}$$

which is symmetric under exchanges $i \leftrightarrow j$, $j \leftrightarrow k$, and $k \leftrightarrow i$ for $\varpi = \pi/2$, $\pi/2 - 3\pi/2$, and $\pi/2 + 3\pi/2$, respectively.

5.1.3   States Having Total Isospin $t = \frac{1}{2}$ and Spin $s = \frac{3}{2}$

In this case, the spin state

$$\left| \left(1 \frac{1}{2} \right) \frac{3}{2} m_s \right\rangle_k^{(S)} \equiv \left| 0 \right\rangle^{(S)} \tag{5.17}$$

is totally symmetric under any pair exchange in spin space. Hence, we have as before

$$|W(\varpi)\rangle_k^{(T)} |0\rangle^{(S)} = \sin \varpi \, |-\rangle_k^{(T)} |0\rangle^{(S)} + \cos \varpi |+\rangle_k^{(T)} |0\rangle^{(S)}$$
$$\equiv \sin \varpi \, \Gamma_{\frac{1}{2}\frac{3}{2}}(M''-) + \cos \varpi \, \Gamma_{\frac{1}{2}\frac{3}{2}}(M''+), \tag{5.18}$$

which is antisymmetric under isospin–spin exchanges of $i \leftrightarrow j$, $j \leftrightarrow k$, and $k \leftrightarrow i$ for $\varpi = \pi/2$, $\pi/2 - 3\pi/2$, and $\pi/2 + 3\pi/2$, respectively. As before the argument

$(M''\pm)$ of the isospin–spin state refers to mixed symmetry for $t = \frac{1}{2}, s = \frac{3}{2}$. Similarly, we also have

$$|W(\varpi - \pi/2)\rangle_k^{(T)}|0\rangle^{(S)} = -\cos\varpi \, |-\rangle_k^{(T)}|0\rangle^{(S)} + \sin\varpi|+\rangle_k^{(T)}|0\rangle^{(S)}$$
$$\equiv -\cos\varpi \, \Gamma_{\frac{1}{2}\frac{3}{2}}(M''-) + \sin\varpi \, \Gamma_{\frac{1}{2}\frac{3}{2}}(M''+), \quad (5.19)$$

which is symmetric under combined isospin–spin exchanges of $i \leftrightarrow j$, $j \leftrightarrow k$, and $k \leftrightarrow i$ for $\varpi = \pi/2$, $\pi/2 - 3\pi/2$, and $\pi/2 + 3\pi/2$, respectively.

5.1.4 States Having Total Isospin $t = \frac{3}{2}$ and Spin $s = \frac{3}{2}$

In this case, we have only one possible state, which is totally symmetric under any pair exchange in isospin–spin space

$$\left|\left(1\frac{1}{2}\right)\frac{3}{2}m_t\right\rangle_k^{(T)}\left|\left(1\frac{1}{2}\right)\frac{3}{2}m_s\right\rangle_k^{(S)} = |0\rangle^{(T)}|0\rangle^{(S)}.$$
$$\equiv \Gamma_{\frac{3}{2}\frac{3}{2}}(S'). \quad (5.20)$$

In this treatment, we have introduced $\Gamma_{ts}^{m_t m_s}(\mathcal{R})$ (in the above the projection quantum numbers m_t, m_s have been suppressed), which are the nine orthonormal representations [namely, $\Gamma_{\frac{1}{2}\frac{1}{2}}(A)$, $\Gamma_{\frac{1}{2}\frac{1}{2}}(S)$, $\Gamma_{\frac{1}{2}\frac{1}{2}}(M-)$, $\Gamma_{\frac{1}{2}\frac{1}{2}}(M+)$, $\Gamma_{\frac{3}{2}\frac{1}{2}}(M'-)$, $\Gamma_{\frac{3}{2}\frac{1}{2}}(M'+)$, $\Gamma_{\frac{1}{2}\frac{3}{2}}(M''-)$, $\Gamma_{\frac{1}{2}\frac{3}{2}}(M''+)$, and $\Gamma_{\frac{3}{2}\frac{3}{2}}(S')$] of isospin–spin states for the trinucleon system.

5.2 Symmetrization of Total Wave Function

To obtain the total wave function having a desired symmetry, we have to combine $\Gamma_{ts}^{m_t m_s}(\mathcal{R})$ with a space wave function having the conjugate symmetry $(\tilde{\mathcal{R}})$. Thus we need to construct space wave function having symmetry $\tilde{\mathcal{R}}$. In Chap. 3, Sect. 5, we saw how one can construct a completely symmetric three-body space wave function, using the kinematic rotation vector. In this section, we discuss how the idea can be generalized to obtain space wave functions having different symmetries.

5.2.1   General Expression for Fully Antisymmetric Wave Function

For the trinucleon, the total wave function should be antisymmetric under exchange of any pair. This can be done by multiplying a particular $\Gamma_{ts}^{m_t m_s}(\mathcal{R})$ by the space wave function of conjugate symmetry. The space wave function for three partitions is given as $\Psi(\vec{z}(\varpi), \vec{z}(\varpi - \pi/2))$, where $\vec{z}(\varpi)$ is given by Eq. (3.43). It is interesting that the three partitions $(ij)k$, $(jk)i$, and $(ki)j$ are obtained for the same values of ϖ, namely, $\pi/2$, $\pi/2 - 2\pi/3$, and $\pi/2 + 2\pi/3$, respectively. The totally symmetric (or antisymmetric) space wave function is obtained by a sum of cyclic permutations of $\Psi(\vec{z}(\varpi), \vec{z}(\varpi - \pi/2))$ which is symmetric (or antisymmetric) under $i \leftrightarrow j$. The latter is done by restricting l_1 values to even (or odd) integers. The sum of cyclic permutations is effected by the operator Σ_0 of Eq. (3.46). For the mixed symmetry states, we first combine a mixed symmetry isospin–spin state with space wave function having conjugate symmetry under $i \leftrightarrow j$, and then perform the cyclic sum (Σ_C). Thus totally antisymmetric trinucleon wave function is given by

$$
\begin{aligned}
\Psi(\xi, \Omega_6) \\
= \Gamma_{\frac{1}{2}\frac{1}{2}}^{m_T m_S}(A) \sum_C \Psi_S^{(+)}\left(\vec{z}(\varpi), \vec{z}\left(\varpi - \frac{\pi}{2}\right)\right) \\
+ \Gamma_{\frac{1}{2}\frac{1}{2}}^{m_T m_S}(S) \sum_C \Psi_A^{(-)}\left(\vec{z}(\varpi), \vec{z}\left(\varpi - \frac{\pi}{2}\right)\right) \\
+ \sum_C \left[\sin 2\varpi \Gamma_{\frac{1}{2}\frac{1}{2}}^{m_T m_S}(M+) - \cos 2\varpi \Gamma_{\frac{1}{2}\frac{1}{2}}^{m_T m_S}(M-)\right] \Psi_M^{(+)}\left(\vec{z}(\varpi), \vec{z}\left(\varpi - \frac{\pi}{2}\right)\right) \\
- \sum_C \left[\sin 2\varpi \Gamma_{\frac{1}{2}\frac{1}{2}}^{m_T m_S}(M-) + \cos 2\varpi \Gamma_{\frac{1}{2}\frac{1}{2}}^{m_T m_S}(M+)\right] \Psi_M^{(-)}\left(\vec{z}(\varpi), \vec{z}\left(\varpi - \frac{\pi}{2}\right)\right) \\
+ \sum_C \left[\sin \varpi \Gamma_{\frac{3}{2}\frac{1}{2}}^{m_T m_S}(M'-) + \cos \varpi \Gamma_{\frac{3}{2}\frac{1}{2}}^{m_T m_S}(M'+)\right] \Psi_{M'}^{(+)}\left(\vec{z}(\varpi), \vec{z}\left(\varpi - \frac{\pi}{2}\right)\right) \\
+ \sum_C \left[-\cos \varpi \Gamma_{\frac{3}{2}\frac{1}{2}}^{m_T m_S}(M'-) + \sin \varpi \Gamma_{\frac{3}{2}\frac{1}{2}}^{m_T m_S}(M'+)\right] \Psi_{M'}^{(-)}\left(\vec{z}(\varpi), \vec{z}\left(\varpi - \frac{\pi}{2}\right)\right) \\
+ \sum_C \left[\sin \varpi \Gamma_{\frac{1}{2}\frac{3}{2}}^{m_T m_S}(M''-) + \cos \varpi \Gamma_{\frac{1}{2}\frac{3}{2}}^{m_T m_S}(M''+)\right] \Psi_{M''}^{(+)}\left(\vec{z}(\varpi), \vec{z}\left(\varpi - \frac{\pi}{2}\right)\right) \\
+ \sum_C \left[-\cos \varpi \Gamma_{\frac{1}{2}\frac{3}{2}}^{m_T m_S}(M''-) + \sin \varpi \Gamma_{\frac{1}{2}\frac{3}{2}}^{m_T m_S}(M''+)\right] \Psi_{M''}^{(-)}\left(\vec{z}(\varpi), \vec{z}\left(\varpi - \frac{\pi}{2}\right)\right) \\
+ \Gamma_{\frac{3}{2}\frac{3}{2}}^{m_T m_S}(S') \sum_C \Psi_{A'}^{(-)}\left(\vec{z}(\varpi), \vec{z}\left(\varpi - \frac{\pi}{2}\right)\right),
\end{aligned}
\tag{5.21}
$$

where $\sum_C$ represents sum over cyclic permutations with $\varpi = \pi/2$, $\pi/2 - 2\pi/3$ and $\pi/2 + 2\pi/3$. The superscript $-(+)$ on Ψ refers to antisymmetry(symmetry) under $i \leftrightarrow j$ [by restricting l_1 to odd(even) values]. The subscript refers to the symmetry component corresponding to the particular values of t and s. To do the indicated cyclic sum in the above equation, we introduce the following operators as in Eq. (3.46):

$$\sum_0 f(\varpi) = \frac{1}{3}\left[f(\varpi) + f\left(\varpi - \frac{2\pi}{3}\right) + f\left(\varpi + \frac{2\pi}{3}\right)\right] \tag{5.22}$$

$$\sum_+ f(\varpi) = \frac{1}{3}\left[2f(\varpi) - f\left(\varpi - \frac{2\pi}{3}\right) - f\left(\varpi + \frac{2\pi}{3}\right)\right] \tag{5.23}$$

$$\sum_- f(\varpi) = \frac{1}{\sqrt{3}}\left[f\left(\varpi - \frac{2\pi}{3}\right) - f\left(\varpi + \frac{2\pi}{3}\right)\right] \tag{5.24}$$

Using these operators, we have the completely antisymmetric total wave function

$$\begin{aligned}
\Psi(\xi, \Omega_6) = {} & \Gamma_{\frac{1}{2}\frac{1}{2}}(A)\phi_S^{(0,+)}(\xi, \Omega_6) + \Gamma_{\frac{1}{2}\frac{1}{2}}(S)\phi_A^{(0,-)}(\xi, \Omega_6) \\
& + \Gamma_{\frac{1}{2}\frac{1}{2}}(M-)\left[\phi_M^{(+,+)}(\xi, \Omega_6) + \phi_M^{(-,-)}(\xi, \Omega_6)\right] \\
& + \Gamma_{\frac{1}{2}\frac{1}{2}}(M+)\left[-\phi_M^{(-,+)}(\xi, \Omega_6) + \phi_M^{(+,-)}(\xi, \Omega_6)\right] \\
& + \Gamma_{\frac{3}{2}\frac{1}{2}}(M'-)\left[\phi_{M'}^{(+,+)}(\xi, \Omega_6) - \phi_{M'}^{(-,-)}(\xi, \Omega_6)\right] \\
& + \Gamma_{\frac{3}{2}\frac{1}{2}}(M'+)\left[\phi_{M'}^{(-,+)}(\xi, \Omega_6) + \phi_{M'}^{(+,-)}(\xi, \Omega_6)\right] \\
& + \Gamma_{\frac{1}{2}\frac{3}{2}}(M''-)\left[\phi_{M''}^{(+,+)}(\xi, \Omega_6) - \phi_{M''}^{(-,-)}(\xi, \Omega_6)\right] \\
& + \Gamma_{\frac{1}{2}\frac{3}{2}}(M''+)\left[\phi_{M''}^{(-,+)}(\xi, \Omega_6) + \phi_{M''}^{(+,-)}(\xi, \Omega_6)\right] \\
& + \Gamma_{\frac{3}{2}\frac{3}{2}}(S)\phi_{A'}^{(0,-)}(\xi, \Omega_6)
\end{aligned} \tag{5.25}$$

The functions $\phi_{\mathcal{R}}^{(\epsilon,\epsilon')}$ are defined as

$$\phi_{\mathcal{R}}^{(0,\epsilon)}(\xi, \Omega_6) = 3\sum_0 \Psi_{\mathcal{R}}^{(\epsilon)}\left(\vec{z}(\varpi), \vec{z}\left(\varpi - \frac{\pi}{2}\right)\right) \qquad (\mathcal{R} = A, S, A') \tag{5.26}$$

$$\phi_{\mathcal{R}}^{(\epsilon',\epsilon)}(\xi, \Omega_6) = \frac{3}{2}\sum_{\epsilon'} \Psi_{\mathcal{R}}^{(\epsilon)}\left(\vec{z}(\varpi), \vec{z}\left(\varpi - \frac{\pi}{2}\right)\right) \tag{5.27}$$

$$(\mathcal{R} = M, M', M'' \text{ and } \epsilon' = + \text{ or } -).$$

5.2.2 Construction of HH for Different Partitions

To apply operators $\sum_{\epsilon}$, one needs HH for different partitions corresponding to the parametric angle ϖ. We note that the hyperangles (denoted by $\Omega_{6,\varpi}$) depend on the partition and thus is a function of the parametric angle ϖ. In our earlier expressions, the hyperangle Ω_6 corresponds to the $(ij)k$ partition, i.e., $\varpi = \pi/2$. Note that the hyperradius ξ is invariant under different permutations. Thus the analytic form of the HH $\mathcal{Y}_{[L]}(\Omega_{6,\varpi})$ for different partitions will be an involved expression in terms of the original hyperangles. On the other hand, each full set of HH $\{\mathcal{Y}_{[L]}(\Omega_{6,\varpi})\}$ for a given ϖ forms a complete set. Hence, a particular HH, $\mathcal{Y}_{[L]}(\Omega_{6,\varpi})$, can be expanded in the complete set $\{\mathcal{Y}(\Omega_6)\}$

$$\mathcal{Y}_{[L]}(\Omega_{6,\varpi}) = \sum_{[L']} \mathcal{C}_{[L][L']}(\varpi)\mathcal{Y}_{[L']}(\Omega_6). \tag{5.28}$$

The expansion coefficients $\mathcal{C}_{[L][L']}(\varpi)$ are called Raynal–Revei coefficients [4].

We already discussed the use of Raynal–Revai coefficients in Chap. 3. Using the coupled basis HH $\mathcal{Y}_{(l_1 l_2)lm, L}(\Omega_6)$, the space wave function $\Psi_{[L]}(\vec{z}(\varpi), \vec{z}(\varpi - \pi/2))$ is expanded in the complete set of coupled HH for the partition corresponding to ϖ in Eq. (3.48). Next the coupled basis HH $\mathcal{Y}_{(l_1 l_2)lm, L}(\Omega_{6,\varpi})$ for the partition ϖ is expanded in the set of original HH according to Eq. (3.49). The Raynal–Revai coefficients $A^{[L]}_{l'_1 l'_2}(\varpi)$ are given by Eqs. (3.50)–(3.52). Note that the value of the hyperangular momentum (grand orbital) quantum number remains unchanged in all partitions, as it is associated with the hyperradius, which is invariant under permutations. Hence, the sum is over $l'_1 l'_2$ only. Using these relations, together with Eqs. (5.22)–(5.28), one can obtain the totally antisymmetric wave function for the trinucleon system.

5.3   Optimal Subset for the Trinucleon

The wave function obtained in the last section involves no approximation. But due to the large number (nine) of symmetry components in Eq. (5.25), the calculations are extremely heavy and computer intensive. Known experimental informations can restrict some of the components. For example, it is known that the ground state of the trinucleon is a $t = \frac{1}{2}$, $J^\pi = \frac{1}{2}^+$ state. Hence, $s = \frac{1}{2}$ and $\frac{3}{2}$ and $l = 0$ and $l = 2$ can contribute, while $l = 1$ has a negligible contribution. Thus the ground state is a mixture of the $l = 0$ space totally symmetric S state, the $l = 0$ space mixed symmetry S' state, and the $l = 2$ space mixed symmetry D state. These correspond, respectively, to the first component, the first terms of third and fourth components and the first terms of seventh and eight components of Eq. (5.25). Hence, the ground state is given by

$$\Psi(\xi, \Omega_6) = \Gamma_{\frac{1}{2}\frac{1}{2}}(A)\phi_S^{(0,+)}(\xi, \Omega_6)$$
$$+ \left[\Gamma_{\frac{1}{2}\frac{1}{2}}(M-)\phi_M^{(+,+)}(\xi, \Omega_6) - \Gamma_{\frac{1}{2}\frac{1}{2}}(M+)\phi_M^{(-,+)}(\xi, \Omega_6) \right]$$
$$+ \left[\Gamma_{\frac{1}{2}\frac{3}{2}}(M''-)\phi_{M''}^{(+,+)}(\xi, \Omega_6) + \Gamma_{\frac{1}{2}\frac{3}{2}}(M''+)\phi_{M''}^{(-,+)}(\xi, \Omega_6) \right]. \tag{5.29}$$

The first, second, and third lines correspond to S, S', and D state, respectively.

A further simplification is possible using the optimal subset approximation, introduced in Chap. 4, Sect. 4.3. Equation (4.36) shows that the optimal subset (OS) depends on the interaction. It is known that the nuclear interaction is a mixture of central and tensor forces, the former being the dominant one. It is consistent with the fact that $l = 0$ and 2 contribute to the ground state. Hence, the S state ($l = 0$) will be the dominant one, which is confirmed by experiments to have a probability of about 90 %. So a representation of space part of the ground state by $\mathcal{Y}_{[0]}(\Omega_6)$ alone will be a good approximation.

Let the optimal subset be denoted by $\{B_k(\Omega_6, s, t)\}$. Clearly, the first member of this set is

$$B_0(\Omega_6, s, t) = \mathcal{Y}_{[0]}(\Omega_6)\Gamma_{\frac{1}{2}\frac{1}{2}}(A) = \pi^{-3/2}\Gamma_{\frac{1}{2}\frac{1}{2}}(A). \tag{5.30}$$

Then by Eq. (4.36), the other members of OS satisfy

$$\int B_k(\Omega_6, s, t)B_{k'}(\Omega_6, s, t)d\Omega_6 = \delta_{kk'}$$

$$\int B_k(\Omega_6, s, t)V(\xi, \Omega_6)B_0(\Omega_6, s, t)d\Omega_6 \neq 0, \tag{5.31}$$

where $V(\xi, \Omega_6)$ is the sum of all interactions, expressed in terms of the hyperspherical variables. Since the net central potential is an even function and is totally symmetric under any pair exchange, it can be expanded in the set of totally symmetric HH with even parity, $\{P_{2K}^{(0)}(\Omega_6)\}$ [see Eq. (5.37)]

$$V(\xi, \Omega_6) = \sum_{K=0}^{\infty} a_{2K} P_{2K}^{(0)}(\Omega_6)V_{2K}(\xi), \tag{5.32}$$

where $V_{2K}(\xi)$ is the potential multipole and a_{2K} is an operator acting on isospin–spin variables, but independent of hyperspherical variables. Condition (5.31) with the first member of OS given by Eq. (5.30) gives the remaining members of the optimal subset (for central interactions) as

$$B_{2K}(\Omega_6, s, t) = C_{2K}P_{2K}^{(0)}\Gamma_{\frac{1}{2}\frac{1}{2}}(A), \tag{5.33}$$

where C_{2K} is a normalization constant. This is true for the central interaction only, for which isospin–spin operator part of a_{2K} is an identity operator and applies to the S state only. For the D state, total angular momentum $\vec{J}$ is obtained by coupling $\vec{s}$ and $\vec{l}$ of the isospin–spin and space wave functions. The isospin–spin wave function is given by [see equations of Sect. 5.1, which define the values of the coefficients $b_s^t(\mathcal{R})$]

$$\Gamma_{t_{ij}s_{ij}}^{m_t m_s}(\mathcal{R}) = \sum_{ts} b_{s_{ij}}^{t_{ij}}(\mathcal{R})|(s_{ij}\frac{1}{2})sm_s\rangle^{(S)}|(t_{ij}\frac{1}{2})tm_t\rangle^{(T)}. \tag{5.34}$$

Since the potential is a sum of central and tensor terms, the expression for a general member of OS, appropriate for any of S, S', and D states, in isospin–spin–space coupled form is given by [1]

$$\left[\Gamma_{ts}(\tilde{\mathcal{R}}) \otimes B^{(\mathcal{R})}_{2K+l}(\Omega_6)\right]_{JM_J} = N^{(\mathcal{R})}_{2K+l} \sum_{l_1 l_2 t_{ij} s_{ij} m m_s} b^{t_{ij}}_{s_{ij}}(\tilde{\mathcal{R}})(-1)^{s-m_J}\hat{i}_1\hat{i}_2\hat{J}$$

$$\begin{pmatrix} l & l_1 & l_2 \\ 0 & 0 & 0 \end{pmatrix}\begin{pmatrix} l & s & J \\ m & m_s & -m_J \end{pmatrix}$$

$$\times\left(\sum_{\mathcal{R}} {}^{(2)}\mathcal{P}^{l_2 l_1}_{2K+l}(\varpi)\right) {}^{(2)}\mathcal{P}^{l_2 l_1}_{2K+l}(\phi)|(l_1 l_2)lm\rangle|(s_{ij}\tfrac{1}{2})sm_s\rangle^{(S)}|(t_{ij}\tfrac{1}{2})tm_t\rangle^{(T)}, \quad (5.35)$$

where $N^{(\mathcal{R})}_{2K+l}$ is a normalization constant. $\tilde{\mathcal{R}}$ represents the conjugate symmetry of $\mathcal{R}$, such that left side of Eq. (5.35) is totally antisymmetric. This subset of HH, called a potential basis, was introduced by Fabre [1] and constitutes the optimal subset for the trinucleon. It was shown by Erens [5] that the OS approximation is a good one for the trinucleon.

In general, l, s, and t are not good quantum numbers, but J, m_J, and m_t are good: $m_t = \frac{1}{2}$ and $m_t = -\frac{1}{2}$ correspond to ^{3}He and ^{3}H nuclei, respectively. Thus the ground state of the trinucleon is a mixture of different components, corresponding to different (l, s, t) values. A component of the trinucleon wave function with orbital angular momentum l, spin s, and isospin t is denoted by $\Psi^{tm_t}_{(ls)Jm_J}(\xi, \Omega_6)$. The $l = 0$ part consists of two components: the totally symmetric S and the mixed symmetry S' states, corresponding to the first two terms of Eq. (5.29). These are expanded in appropriate optimal subsets

$$\Psi^{\frac{1}{2}m_t}_{(0\frac{1}{2})\frac{1}{2}\frac{1}{2}}(\xi, \Omega_6) = \Gamma^{m_t\frac{1}{2}}_{\frac{1}{2}\frac{1}{2}}(A)\xi^{-5/2}\sum_K P^{(0)}_{2K}(\Omega_6)u^{S\frac{1}{2}\frac{1}{2}}_{2K}(\xi)$$

$$+ \xi^{-5/2}\sum_K 2^{-1/2}\left[\Gamma^{m_t\frac{1}{2}}_{\frac{1}{2}\frac{1}{2}}(M-)P^{(+)}_{2K}(\Omega_6) - \Gamma^{m_t\frac{1}{2}}_{\frac{1}{2}\frac{1}{2}}(M+)P^{(-)}_{2K}(\Omega_6)\right]u^{S'\frac{1}{2}\frac{1}{2}}_{2K}(\xi),$$

$$(5.36)$$

where $u^{St=\frac{1}{2}s=\frac{1}{2}}_{2K}(\xi)$ and $u^{S't=\frac{1}{2}s=\frac{1}{2}}_{2K}(\xi)$ are hyperradial partial waves for the S and S' states, respectively. Hyperangular parts of OS elements with different symmetries ($\epsilon = 0, +$ and $-$) for the S and S' states are given by

$$P^{(\epsilon)}_{2K}(\Omega_6) \equiv B^{(S,\epsilon)}_{2K}(\Omega_6)$$

$$= N^{(\epsilon)}_{2K}\sum_{l_1=0}^{K} {}^{(\epsilon)}F^{l_1 l_1}_{2K}(\frac{\pi}{2}) {}^{(2)}\mathcal{P}^{l_1 l_1}_{2K}(\phi)\sum_{m_1=-l_1}^{l_1} Y_{l_1 m_1}(\hat{\xi}_1)Y^*_{l_1 m_1}(\hat{\xi}_2),$$

$$\text{(with } \epsilon = 0, + \text{ and } -). \quad (5.37)$$

The F functions introduce the desired symmetry and is given by

$$^{(\epsilon)}F_{2K}^{l_1l_1}(\varpi) = \sum_{\epsilon} {}^{(2)}\mathcal{P}_{2K}^{l_1l_1}(\varpi), \tag{5.38}$$

and the normalizing constant is

$$[N_{2K}^{(\epsilon)}]^{-2} = \sum_{l_1=0}^{K}(2l_1 + 1)\left[{}^{(\epsilon)}F_{2K}^{l_1l_1}(\varpi)\right]^2. \tag{5.39}$$

In a similar fashion, we can expand the complete D state wave function [corresponding to the third term of Eq. (5.29)] in an appropriate OS

$$\Psi_{(2\frac{3}{2})\frac{1}{2}\frac{1}{2}}^{\frac{1}{2}m_t}(\xi, \Omega_6) = \xi^{-5/2}\sum_{K}2^{-1/2}\left[D_{2K+2}^{(+)}(\Omega_6)\left|\left(0\frac{1}{2}\right)\frac{1}{2}m_t\right\rangle^{(T)}\right.$$

$$+ \left.D_{2K+2}^{(-)}(\Omega_6)\left|\left(1\frac{1}{2}\right)\frac{1}{2}m_t\right\rangle^{(T)}\right]u_{2K+2}^{D\frac{1}{2}\frac{3}{2}}(\xi), \tag{5.40}$$

where $u_{2K+2}^{D\frac{1}{2}\frac{3}{2}}(\xi)$ is the D state partial wave and

$$D_{2K+2}^{(\epsilon)}(\Omega_6) = B_{2K+2}^{(D,\epsilon)}$$

$$= N_{2K+2}^{(D,\epsilon)}\sum_{l_1l_2}\hat{l}_1\hat{l}_2\begin{pmatrix}2 & l_1 & l_2 \\ 0 & 0 & 0\end{pmatrix}$$

$$\times {}^{(D,\epsilon)}F_{2K+2}^{l_2l_1}(\pi/2)\,{}^{(2)}\mathcal{P}_{2K+2}^{l_2l_1}(\phi)\left\langle\hat{\xi}_1\hat{\xi}_2\left|(l_1l_2)2;\left(1\frac{1}{2}\right)\frac{3}{2};\frac{1}{2}\frac{1}{2}\right\rangle\right. \tag{5.41}$$

with the normalization constant given by

$$[N_{2K+2}^{(D,\epsilon)}]^{-2} = \sum_{l_1l_2}(2l_1 + 1)(2l_2 + 1)\begin{pmatrix}2 & l_1 & l_2 \\ 0 & 0 & 0\end{pmatrix}\left[{}^{(D,\epsilon)}F_{2K+2}^{l_1l_2}(\pi/2)\right]^2 \tag{5.42}$$

and

$$^{(D,\epsilon)}F_{2K+2}^{l_2l_1}(\varpi) = \sum_{\epsilon}{}^{(2)}\mathcal{P}_{2K+2}^{l_2l_1}(\varpi). \tag{5.43}$$

In the above $\sum_{\epsilon}$ is given by Eqs. (5.22)–(5.24).

5.4 Calculation of Potential Matrix Element: GSC

Since the general elements of the OS are given by Eq. (5.35), the matrix elements of
the potential are given by

$$
V_{\alpha,\alpha'}(\xi) = \langle \left[\Gamma_{ts}(\tilde{\mathcal{R}}) \otimes B^{(\mathcal{R})}_{2K+l}(\Omega_6) \right]_{JM_J} \Big| V(\xi,\Omega_6) \Big| \left[\Gamma_{t's'}(\tilde{\mathcal{R}}') \otimes B^{(\mathcal{R}')}_{2K'+l'}(\Omega_6) \right]_{JM_J} \rangle,
$$
(5.44)

where $\alpha = \{t, s, K, l, \mathcal{R}\}$ is an abbreviation of relevant quantum numbers. Since the
OS is totally antisymmetric under any pair exchange, the potential matrix element
(PME) of the total potential is just the PME of any one pair [say (ij)-pair] times the
number of pairs. Hence,

$$
V_{\alpha,\alpha'}(\xi) = 3 \langle \left[\Gamma_{ts}(\tilde{\mathcal{R}}) \otimes B^{(\mathcal{R})}_{2K+l}(\Omega_6) \right]_{JM_J} \Big| V(\vec{r}_{ij})) \Big| \left[\Gamma_{t's'}(\tilde{\mathcal{R}}') \otimes B^{(\mathcal{R}')}_{2K'+l'}(\Omega_6) \right]_{JM_J} \rangle.
$$
(5.45)

Now the nucleon–nucleon potential is a sum of central and tensor potentials. Hence,
(ij)-pair potential can be expanded in HH as

$$
V(\vec{r}_{ij}) = V(\vec{\xi}_1)
$$
$$
= \sum_{l''(=0,2)m_l''} a_{ij}^{l''m_{l''}}(\vec{t}_i, \vec{t}_j, \vec{s}_i, \vec{s}_j) \sum_{K''} V_{2K''+l''}(\xi)\,{}^{(2)}\mathcal{P}^{0,l''}_{2K''+l''}(\phi) Y_{l''m_{l''}}(\vartheta_1, \varphi_1), \quad (5.46)
$$

where $l'' = 0$ and 2 correspond to the central and tensor terms and $a_{ij}^{l''m_{l''}}(\vec{t}_i, \vec{t}_j, \vec{s}_i, \vec{s}_j)$
is the operator acting on isospin–spin part of the wave function. The spin operators
$\vec{\sigma}_i$ and $\vec{\sigma}_j$ (vectors in 3D space) of $a_{ij}^{l''m_{l''}}(\vec{t}_i, \vec{t}_j, \vec{s}_i, \vec{s}_j)$ and $Y_{l''m_{l''}}(\vartheta_1, \varphi_1)$ (a spherical
tensor of rank l'' in 3D space) are to be coupled to a scaler, so that the potential
is invariant under combined spin–space rotations in 3D space [1]. The potential
multipole (PM) $V_{2K''+l''}(\xi)$ can be calculated easily by a single one-dimensional
integral over ϕ.

From Eqs. (5.45) and (5.46), we see that the PME will be a sum of terms containing
product of $\langle \Gamma_{ts}(\tilde{\mathcal{R}}) | a_{ij}^{l''m_{l''}}(\vec{t}_i, \vec{t}_j, \vec{s}_i, \vec{s}_j) | \Gamma_{t's'}(\tilde{\mathcal{R}}') \rangle$ with a geometrical structure coeffi-
cient (GSC) [see Chap. 3 Sect. 6], involving the 3P-coefficients $\langle {}^{(2)}\mathcal{P}^{l_2l_1}_{2K+l} | {}^{(2)}\mathcal{P}^{0\,l''}_{2K''+l''} | {}^{(2)}\mathcal{P}^{l'_2l'_1}_{2K'+l'} \rangle$. These can be calculated by the method of linear inhomogeneous equa-
tions proposed by De and Das [6, 7] and discussed in Chap. 3 for the simple S
state of the trinucleon. We briefly recapitulate the method. Using the definition of
${}^{(2)}\mathcal{P}$-function [Eq. (3.26)], we have

$$
\langle {}^{(2)}\mathcal{P}^{l_2l_1}_{2K+l} \Big| {}^{(2)}\mathcal{P}^{0l''}_{2K''+l''} \Big| {}^{(2)}\mathcal{P}^{l'_2l'_1}_{2K'+l'} \rangle
$$
$$
= \int_0^{\frac{\pi}{2}} {}^{(2)}\mathcal{P}^{l_2l_1}_{2K+l}(\phi)\,{}^{(2)}\mathcal{P}^{0l''}_{2K''+l''}(\phi)\,{}^{(2)}\mathcal{P}^{l'_2l'_1}_{2K'+l'}(\phi)\,\sin^2\phi\,\cos^2\phi\,d\phi
$$

$$
= N_{2K+l}^{l_2 l_1} N_{2K''+l''}^{0 l''} N_{2K'+l'}^{l_2' l_1'} 2^{-(\tilde{n}+3)} \int_{-1}^{1} (1-x)^{(l_1+l_1'+l''+1)/2} (1+x)^{(l_2+l_2'+1)/2}
$$

$$
\times \ P_{K+(l-l_1-l_2)/2}^{l_1+\frac{1}{2}, l_2+\frac{1}{2}}(x) P_{K''}^{l''+\frac{1}{2}, \frac{1}{2}}(x) P_{K'+(l'-l_1'-l_2')/2}^{l_1'+\frac{1}{2}, l_2'+\frac{1}{2}}(x) dx, \tag{5.47}
$$

where $\tilde{n} = (l_1 + l_1' + l_2 + l_2' + l'')/2$. Note that this is an integer, since $(l_1 + l_2)$ and $(l_1' + l_2')$ are even integers (wave function has even parity) and $l'' = 0$ or 2 (corresponding to central and tensor interactions respectively).

Selection Rule

We can easily see that the 3P coefficient given by Eq. (5.47) vanishes, unless a selection rule is satisfied. Since $\tilde{\mathcal{P}}_L^{l_2 l_1} = \xi^L \ {}^{(2)}\mathcal{P}_L^{l_2 l_1}(\phi)$ is a homogeneous polynomial of degree L in Cartesian components of the Jacobi vectors, the product $\tilde{\mathcal{P}}_{2K+l}^{l_2 l_1} \ \tilde{\mathcal{P}}_{2K'+l'}^{l_2' l_1'}$ is a homogeneous polynomial of degree $(2K + l + 2K' + l')$. Hence, it can be expanded in a series of polynomials of lower degree, multiplied by ξ raised to the difference of the degrees [7]

$$
\tilde{\mathcal{P}}_{2K+l}^{l_2 l_1} \ \tilde{\mathcal{P}}_{2K'+l'}^{l_2' l_1'} = \sum_{K''} c_{K''} \tilde{\mathcal{P}}_{2K''+l''}^{l_2'' l_1''} \ \xi^{\{2K+l+2K'+l'-(2K''+l'')\}}, \tag{5.48}
$$

where $c_{K''}$ are constants. Since the polynomials must be regular at the origin, the exponent of ξ on the right side must be a positive definite integer. Hence, $2K + l + 2K' + l' \geq 2K'' + l''$. Permutations of the three ${}^{(2)}\mathcal{P}$ functions give similar relations. Combining these we get the *triangle inequality* for a triangle having sides $(2K + l)$, $(2K' + l')$, and $(2K'' + l'')$

$$
|(2K + l) - (2K' + l')| \leq (2K'' + l'') \leq \{(2K + l) + (2K' + l')\}. \tag{5.49}
$$

This condition restricts the allowed values of K'' for given values of K, K', l, l', and l''. The corresponding GSC vanishes unless this condition is satisfied. Expressing Eq. (5.48) in terms of ${}^{(2)}\mathcal{P}$ functions, multiplying both sides by a third ${}^{(2)}\mathcal{P}$ function, integrating, and using their orthonormal property [Eq. (3.25)], we see that $\left\langle {}^{(2)}\mathcal{P}_{2K+l}^{l_2 l_1} \middle| {}^{(2)}\mathcal{P}_{2K''+l''}^{0 l''} \middle| {}^{(2)}\mathcal{P}_{2K'+l'}^{l_2' l_1'} \right\rangle$ will be nonvanishing only for values of K'' given by Eq. (5.49). We will derive specific selection rules for K'' in specific cases, corresponding to appropriate values of l, l' and l''.

In the following subsections, we discuss explicitly the coupling of S, S', and D states through central and tensor forces. Coupling of two S states in the $(ij)k$ partition through central (ij)-pair interaction was discussed in Chap. 3, Sect. 6. There we discussed how a whole set of GSCs can be calculated by solving a single set of linear inhomogeneous equations (LIE), resulting in a very fast computation with very high precision. The method can easily be generalized for components (in the OS) of the totally antisymmetric state of the trinucleon coupled through central or tensor interactions.

5.4.1 Coupling Among S and S' States Through Central interactions

Both S and S' states have $l = 0$. They can be coupled through central interaction. The OS expansion of S and S' states is given by Eq. (5.36). The GSC necessary for coupling two S states, two S' states, or between S and S' states is given by [7]

$$
\langle K, \epsilon | K'', l'' = 0 | K', \epsilon' \rangle
$$

$$
= \frac{\sqrt{\pi}}{4} (K'' + 1) N_{2K}^{(\epsilon)} N_{2K'}^{(\epsilon')}
$$

$$
\times \sum_{l=0}^{\min(K, K')} (2l + 1)^{(\epsilon)} F_{2K}^{ll} \left(\frac{\pi}{2} \right) {}^{(\epsilon')} F_{2K'}^{ll} \left(\frac{\pi}{2} \right) \left\langle {}^{(2)} \mathcal{P}_{2K}^{ll} \left| {}^{(2)} \mathcal{P}_{2K''}^{00} \right| {}^{(2)} \mathcal{P}_{2K'}^{ll} \right\rangle, \quad (5.50)
$$

where ϵ and ϵ' correspond to superscripts $= 0, +$ and $-$ of Eq. (5.36) and F-functions and normalization constants are given by Eqs. (5.38) and (5.39), respectively. The selection rule for K'' values is obtained from Eq. (5.49), by setting $l = l' = l'' = 0$

$$
\langle K, \epsilon | K'', l'' = 0 | K', \epsilon' \rangle = 0 \ \text{unless} \ |K - K'| \le K'' \le (K + K'). \quad (5.51)
$$

A double sum was derived by Fabre [1] for the evaluation of the GSC. However, the double sum was an alternating series, whose terms involve ratios of gamma function of large arguments. Thus it involved large numerical errors for large values of K and K'. Moreover, the calculations have to be performed for each combination of K, K', and K''. Instead, we derive a set of linear inhomogeneous equations (LIE) using the completeness property of the Jacobi polynomials, Eq. (3.56). Multiplying both sides of Eq. (5.50) by $K''! P_{K''}^{\frac{1}{2}, \frac{1}{2}}(y) / \Gamma(K'' + \frac{3}{2})$, summing over K'', and using Eq. (3.56) and (5.47), we get

$$
\sum_{K''} K''! \, P_{K''}^{\frac{1}{2}, \frac{1}{2}}(y) \, \langle K, \epsilon | K'', l'' = 0 | K', \epsilon' \rangle / \Gamma \left(K'' + \frac{3}{2} \right) = \frac{\sqrt{\pi}}{8} N_{2K}^{(\epsilon)} N_{2K'}^{(\epsilon')}
$$

$$
\times \sum_{l=0}^{\min(K, K')} (2l + 1) \, {}^{(\epsilon)} F_{2K}^{ll} \left(\frac{\pi}{2} \right) {}^{(\epsilon')} F_{2K'}^{ll} \left(\frac{\pi}{2} \right) \, 2^{-2l} (1 - y)^l (1 + y)^l
$$

$$
\times N_{2K}^{l,l} \, N_{2K'}^{l,l} \, P_{K-l}^{l+\frac{1}{2}, l+\frac{1}{2}}(y) \, P_{K'-l}^{l+\frac{1}{2}, l+\frac{1}{2}}(y), \quad (5.52)
$$

where $N_L^{l_2, l_1}$ is the normalization constant of ${}^{(2)} \mathcal{P}_L^{l_2, l_1}(\phi)$, given by Eq. (3.26). Equation (5.52) is a *finite* set of LIE due to the selection rule Eq. (5.51). Let $n_{KK'}$ be the number of nonvanishing GSCs for given values of K and K'. Since Eq. (5.52) is valid for any value of y in the interval $-1 \le y \le 1$, we can choose $n_{KK'}$ different values of y. Then Eq. (5.52) is a set of $n_{KK'}$ equations for the $n_{KK'}$ unknown GSC's for given values of K and K'. Solving this set of LIE, we get *all the required GSCs* for given values of K and K', in a single step. The LIE can be solved by a very fast and accurate algorithm. One can also calculate the GSC directly by doing the integral in Eq. (5.47) numerically. It was found by De and Das [7] that the LIE method is both fast and accurate, compared with the double sum or the integral methods.

Setting $y = 1$ in Eq. (5.52), we obtain a simple *sum rule* for the GSCs

$$\sum_{K''=|K-K'|}^{K+K'} \langle K, \epsilon | K'', l'' = 0 | K', \epsilon' \rangle$$

$$= N_{2K}^{(\epsilon)} N_{2K'}^{(\epsilon')} {}^{(\epsilon)}F_{2K}^{00}\left(\frac{\pi}{2}\right) {}^{(\epsilon')}F_{2K'}^{00}\left(\frac{\pi}{2}\right) (K+1)(K'+1). \tag{5.53}$$

This relation can be used to test the accuracy of calculated GSCs.

5.4.2 Coupling Between S and D State Through Tensor Interaction

The GSC for this coupling is given by [7]

$${}_D\langle K, + | K'', T | K', 0 \rangle_S = \frac{\sqrt{2\pi}}{64} N_{2K+2}^{(D,+)} N_{2K'}^{(0)} {}^{(2)}\mathcal{P}_{2K''+2}^{2,0}(0)$$

$$\times \sum_{l_1,l_2} (2l_1 + 1)(2l_2 + 1) \begin{pmatrix} 2 & l_1 & l_2 \\ 0 & 0 & 0 \end{pmatrix}^2 {}^{(0)}F_{2K'}^{l_1 l_1}\left(\frac{\pi}{2}\right)$$

$$\times {}^{(D,+)}F_{2K+2}^{l_2 l_1}\left(\frac{\pi}{2}\right) \left(\left| {}^{(2)}\mathcal{P}_{2K+2}^{l_2,l_1} \right| {}^{(2)}\mathcal{P}_{2K''+2}^{2,0} \right|^{(2)} \mathcal{P}_{2K'}^{l_1,l_1} \right). \tag{5.54}$$

Here K'',T corresponds to the K'' multipole of the tensor force ($l'' = 2$). The normalization constants and F-functions are given by Eqs. (5.38), (5.39), (5.42), and (5.43). As before, multiplying both sides of Eq. (5.54) by $P_{K''}^{\frac{1}{2},\frac{5}{2}}(y)/P_{K''}^{\frac{1}{2},\frac{5}{2}}(1)$, summing over K'', and using the completeness property of Jacobi polynomials, Eq. (3.56), we have

$$\sum_{K''} [P_{K''}^{\frac{1}{2},\frac{5}{2}}(y)/P_{K''}^{\frac{1}{2},\frac{5}{2}}(1)] \, {}_D\langle K, + | K'', T | K', 0 \rangle_S = \frac{\sqrt{2\pi}}{32} N_{2K+2}^{(D,+)} N_{2K'}^{(0)}$$

$$\times \sum_{l_1,l_2} (2l_1 + 1)(2l_2 + 1) N_{2K+2}^{l_2,l_1} N_{2K'}^{l_1,l_1} \begin{pmatrix} 2 & l_1 & l_2 \\ 0 & 0 & 0 \end{pmatrix}^2$$

$$\times {}^{(0)}F_{2K'}^{l_1 l_1}(\frac{\pi}{2}) \, {}^{(D,+)}F_{2K+2}^{l_2 l_1}\left(\frac{\pi}{2}\right) (1+y)^{(l_1+l_2)/2-1}(1-y)^{l_1} 2^{-(3l_1+l_2)/2}$$

$$\times P_{K'-l_1}^{l_1+\frac{1}{2},l_1+\frac{1}{2}}(y) P_{(2K+2-l_2-l_1)/2}^{l_1+\frac{1}{2},l_2+\frac{1}{2}}(y). \tag{5.55}$$

Selection rule: From Eq. (5.49), setting $l = 2, l' = 0$ (corresponding to D and S states respectively) and $l'' = 2$ (for tensor force), we get

$$_D\langle K, +|K'', T|K', 0\rangle_S = 0$$

unless $$\max\{K - K', K' - K - 2, 0\} \leq K'' \leq (K + K') \qquad (5.56)$$

Thus the number of terms on left side of Eq. (5.55) is a finite one and we can follow LIE method by choosing $n_{KK'}$ different values of y in the interval $[-1, 1]$, where $n_{KK'}$ is the number of nonvanishing GSCs, according to Eq. (5.56).

As before we can obtain a *sum rule* for these GSCs, by setting $y = 1$ in Eq. (5.55) (note that l_1 must vanish due to the factor $(1 - y)^{l_1}$)

$$\sum_{K''=\max(K-K',K'-K-2,0)}^{K+K'} {}_D\langle K, +|K'', T|K', 0\rangle_S$$

$$= \frac{1}{\sqrt{2\pi}} N_{2K+2}^{(D,+)} N_{2K'}^{(0)} \, {}^{(D,+)}F_{2K+2}^{2,0}\left(\frac{\pi}{2}\right) \, {}^{(0)}F_{2K'}^{0,0}\left(\frac{\pi}{2}\right)$$

$$\times (K + 2)(K' + 1)\left[\frac{(K + 1)(K + 3)}{(2K + 3)(2K + 5)}\right]^{\frac{1}{2}}, \qquad (5.57)$$

which provides precision of numerically calculated GSCs.

5.4.3 Coupling Between S' and D States Through Tensor Interaction

The GSC for such a coupling is given by

$$_D\langle K, \epsilon|K'', T|K', \epsilon'\rangle_{S'} = (-\epsilon)\delta_{\epsilon\epsilon'}\frac{\sqrt{\pi}}{64} N_{2K+2}^{(D,\epsilon)} N_{2K'}^{(\epsilon)} \, {}^{(2)}\mathcal{P}_{2K''+2}^{2,0}(0)$$

$$\times \sum_{l_1,l_2 \text{ (parity of } \epsilon)} (2l_1 + 1)(2l_2 + 1)\begin{pmatrix} 2 & l_1 & l_2 \\ 0 & 0 & 0 \end{pmatrix}^2$$

$$\times {}^{(D,\epsilon)}F_{2K+2}^{l_2 l_1}\left(\frac{\pi}{2}\right) \, {}^{(\epsilon)}F_{2K'}^{l_1 l_1}\left(\frac{\pi}{2}\right)$$

$$\times \langle {}^{(2)}\mathcal{P}_{2K+2}^{l_2,l_1}|{}^{(2)}\mathcal{P}_{2K''+2}^{2,0}|{}^{(2)}\mathcal{P}_{2K'}^{l_1,l_1}\rangle. \qquad (5.58)$$

Note that the symmetry under P_{12} (i.e. ϵ) for the D and S' states must be the same, since the interaction (including tensor) conserves parity. To obtain the LIE for calculation of the GSCs, we multiply both sides of Eq. (5.58) by $P_{K''}^{\frac{1}{2},\frac{5}{2}}(y)/P_{K''}^{\frac{1}{2},\frac{5}{2}}(1)$, sum over K'', and use the completeness property of Jacobi polynomials, Eq. (3.56), and get

$$\sum_{K''} \left[P_{K''}^{\frac{1}{2},\frac{5}{2}}(y)/P_{K''}^{\frac{1}{2},\frac{5}{2}}(1) \right] {}_{\mathrm{D}}\langle K, \epsilon | K'', \mathrm{T} | K', \epsilon' \rangle_{\mathrm{S'}} = (-\epsilon)\delta_{\epsilon\epsilon'} \frac{\sqrt{\pi}}{32} N_{2K+2}^{(\mathrm{D},\epsilon)} N_{2K'}^{(\epsilon)}$$

$$\times \sum_{l_1,l_2 \,(\text{parity of } \epsilon)} (2l_1+1)(2l_2+1) N_{2K+2}^{l_2,l_1} N_{2K'}^{l_1,l_1} \begin{pmatrix} 2 & l_1 & l_2 \\ 0 & 0 & 0 \end{pmatrix}^2$$

$$\times {}^{(\mathrm{D},\epsilon)} F_{2K+2}^{l_2 l_1}\left(\frac{\pi}{2}\right) {}^{(\epsilon)} F_{2K'}^{l_1 l_1}\left(\frac{\pi}{2}\right) (1+y)^{(l_1+l_2)/2-1}(1-y)^{l_1} 2^{-(3l_1+l_2)/2}$$

$$\times P_{K'-l_1}^{l_1+\frac{1}{2},l_1+\frac{1}{2}}(y) P_{(2K+2-l_2-l_1)/2}^{l_1+\frac{1}{2},l_2+\frac{1}{2}}(y). \tag{5.59}$$

Selection rule: Since $l = 2$, $l'' = 2$, and $l' = 0$ are same as in the case of S and D state coupling through tensor interaction, the selection rule will be the same and is given by Eq. (5.56). Once again, the number of nonvanishing GSCs for S'-D coupling is finite and the set of LIE can be used with the appropriate number of y values in Eq. (5.59).

Sum rule: As in the previous cases, we obtain a sum rule by setting $y = 1$ in Eq. (5.59)

$$\sum_{K''=\max(K-K',K'-K-2,0)}^{K+K'} {}_{\mathrm{D}}\langle K, + | K'', \mathrm{T} | K', + \rangle_{\mathrm{S'}}$$

$$= \frac{1}{2\sqrt{\pi}} N_{2K+2}^{(\mathrm{D},+)} N_{2K'}^{(+)} {}^{(\mathrm{D},+)} F_{2K+2}^{2,0}\left(\frac{\pi}{2}\right) {}^{(+)} F_{2K'}^{0,0}\left(\frac{\pi}{2}\right)$$

$$\times (K+2)(K'+1)\left[\frac{(K+1)(K+3)}{(2K+3)(2K+5)}\right]^{\frac{1}{2}},$$

and

$$\sum_{K''=\max(K-K',K'-K-2,0)}^{K+K'} {}_{\mathrm{D}}\langle K, - | K'', \mathrm{T} | K', - \rangle_{\mathrm{S'}} = 0. \tag{5.60}$$

The sum vanishes for $\epsilon = -$, since in this case the value of l_1 must be odd [note that ${}^{(-)} F_{2K}^{l_1,l_1}(\pi/2)$ vanishes for even l_1; see Eqs. (5.36–5.38), (5.24) and (3.26)], for which $(1-y)^{l_1} = 0$ for $y = 1$. For $\epsilon = +$, l_1 must be even and $l_1 = 0$ contributes nonvanishingly.

5.4.4 Coupling Between Two D States Through Central Forces

Two D states can couple either through central or through tensor interactions. The GSC for the former is given as

$$_D\langle K, \epsilon | K'', C | K', \epsilon' \rangle_D = \delta_{\epsilon\epsilon'} 2^{-7} N_{2K+2}^{(D,\epsilon)} N_{2K'+2}^{(D,\epsilon)} {}^{(2)}\mathcal{P}_{2K''}^{0,0}(0)$$

$$\times \sum_{l_1,l_2} (2l_1 + 1)(2l_2 + 1) \begin{pmatrix} 2 & l_1 & l_2 \\ 0 & 0 & 0 \end{pmatrix}^2$$

$$\times {}^{(D,\epsilon)} F_{2K+2}^{l_2 l_1}\left(\frac{\pi}{2}\right) {}^{(D,\epsilon)} F_{2K'+2}^{l_2 l_1}\left(\frac{\pi}{2}\right)$$

$$\times \langle {}^{(2)}\mathcal{P}_{2K+2}^{l_2,l_1} | {}^{(2)}\mathcal{P}_{2K''}^{0,0} | {}^{(2)}\mathcal{P}_{2K'+2}^{l_2,l_1} \rangle. \tag{5.61}$$

As before, multiplying both sides of Eq. (5.61) by $P_{K''}^{\frac{1}{2},\frac{1}{2}}(y)/P_{K''}^{\frac{1}{2},\frac{1}{2}}(1)$, summing over K'', and using the completeness property of Jacobi polynomials, Eq. (3.56), we get the set of LIE

$$\sum_{K''} \left[P_{K''}^{\frac{1}{2},\frac{1}{2}}(y)/P_{K''}^{\frac{1}{2},\frac{1}{2}}(1) \right] {}_D\langle K, \epsilon | K'', C | K', \epsilon' \rangle_D = \delta_{\epsilon\epsilon'} 2^{-7} N_{2K+2}^{(D,\epsilon)} N_{2K'+2}^{(D,\epsilon)}$$

$$\times \sum_{l_1,l_2} (2l_1 + 1)(2l_2 + 1) \begin{pmatrix} 2 & l_1 & l_2 \\ 0 & 0 & 0 \end{pmatrix}^2 {}^{(D,\epsilon)} F_{2K+2}^{l_2 l_1}\left(\frac{\pi}{2}\right) {}^{(D,\epsilon)} F_{2K'+2}^{l_2 l_1}\left(\frac{\pi}{2}\right)$$

$$\times 2^{-(l_1+l_2)} N_{2K+2}^{l_2,l_1} N_{2K'+2}^{l_2,l_1}(1 - y)^{l_1}(1 + y)^{l_2} P_n^{l_1+\frac{1}{2},l_2+\frac{1}{2}}(y) P_{n'}^{l_1+\frac{1}{2},l_2+\frac{1}{2}}(y), \tag{5.62}$$

where $n = (2K + 2 - l_1 - l_2)/2$ and $n' = (2K' + 2 - l_1 - l_2)/2$.

Selection rule: Since $l = 2$, $l' = 2$, and $l'' = 0$, the selection rule following Eq. (5.49) is

$$|K - K'| \leq K'' \leq (K + K' + 2). \tag{5.63}$$

Once again, the number of nonvanishing GSCs is finite and the set of LIE can be solved with the appropriate number of y values in Eq. (5.62).

Sum rule: Setting $y = 1$ in Eq. (5.62), we get a sum rule for this case as

$$\sum_{K''=|K-K'|}^{K+K'+2} {}_D\langle K, + | K'', C | K', + \rangle_D = (2\pi)^{-1} N_{2K+2}^{(D,+)} N_{2K'+2}^{(D,+)}$$

$$\times {}^{(D,+)} F_{2K+2}^{2,0}\left(\frac{\pi}{2}\right) {}^{(D,+)} F_{2K'+2}^{2,0}\left(\frac{\pi}{2}\right)$$

$$\times (K + 2)(K' + 2) \left[\frac{(K + 1)(K + 3)}{(2K + 3)(2K + 5)} \frac{(K' + 1)(K' + 3)}{(2K' + 3)(2K' + 5)} \right]^{\frac{1}{2}},$$

and

$$\sum_{K''=|K-K'|}^{K+K'+2} {}_D\langle K, - | K'', C | K', - \rangle_D = 0. \tag{5.64}$$

As in Eq. (5.60), the sum for $\epsilon = -$ vanishes.

5.4.5 Coupling Between Two D States Through Tensor Interaction

The GSC for such a coupling is given by

$$
\begin{aligned}
{}_D\langle K, \epsilon | K'', T | K', \epsilon' \rangle_D &= -\delta_{\epsilon\epsilon'} 2^{-7} \sqrt{\frac{35}{2}} N_{2K+2}^{(D,\epsilon)} N_{2K'+2}^{(D,\epsilon)} \, {}^{(2)}P_{2K''+2}^{2,0}(0) \\
&\quad \times \sum_{l_1, l_2, l_2'} (2l_1 + 1)(2l_2 + 1)(2l_2' + 1) \begin{pmatrix} 2 & l_1 & l_2 \\ 0 & 0 & 0 \end{pmatrix} \begin{pmatrix} 2 & l_1 & l_2' \\ 0 & 0 & 0 \end{pmatrix} \\
&\quad \times \begin{pmatrix} 2 & l_2 & l_2' \\ 0 & 0 & 0 \end{pmatrix} \begin{bmatrix} 2 & 2 & 2 \\ l_2 & l_1 & l_2' \end{bmatrix} {}^{(D,\epsilon)}F_{2K+2}^{l_2 l_1} \left(\frac{\pi}{2}\right) {}^{(D,\epsilon)}F_{2K'+2}^{l_2' l_1} \left(\frac{\pi}{2}\right) \\
&\quad \times {}^{(2)}P_{2K+2}^{l_2, l_1} |{}^{(2)}P_{2K''}^{2,0}|{}^{(2)}P_{2K'+2}^{l_2', l_1},
\end{aligned}
$$

(5.65)

where

$$
\begin{bmatrix} j_1 & j_2 & j_{12} \\ j & j_3 & j_{13} \end{bmatrix}
$$

is a 6-j symbol, where $\vec{j} = \vec{j}_1 + \vec{j}_2 + \vec{j}_3$ is the resultant of three angular momenta and $\vec{j}_{12} = \vec{j}_1 + \vec{j}_2$ and $\vec{j}_{13} = \vec{j}_1 + \vec{j}_3$ are the angular momenta of intermediate couplings. To obtain the set of LIE for computation of these GSCs, we multiply both sides of Eq. (5.65) by $P_{K''}^{\frac{1}{2}, \frac{5}{2}}(y)/P_{K''}^{\frac{1}{2}, \frac{5}{2}}(1)$, sum over K'', and use the completeness property of Jacobi polynomials, Eq. (3.56), to get

$$
\begin{aligned}
\sum_{K''} [P_{K''}^{\frac{1}{2}, \frac{5}{2}}(y)/P_{K''}^{\frac{1}{2}, \frac{5}{2}}(1)] \, {}_D\langle K, \epsilon | K'', T | K', \epsilon' \rangle_D \\
= -\delta_{\epsilon\epsilon'} 2^{-6} \sqrt{\frac{35}{2}} \, N_{2K+2}^{(D,\epsilon)} N_{2K'+2}^{(D,\epsilon)} \sum_{l_1, l_2, l_2'} (2l_1 + 1)(2l_2 + 1)(2l_2' + 1) \\
\times \begin{pmatrix} 2 & l_1 & l_2 \\ 0 & 0 & 0 \end{pmatrix} \begin{pmatrix} 2 & l_1 & l_2' \\ 0 & 0 & 0 \end{pmatrix} \begin{pmatrix} 2 & l_2 & l_2' \\ 0 & 0 & 0 \end{pmatrix} \begin{bmatrix} 2 & 2 & 2 \\ l_2 & l_1 & l_2' \end{bmatrix} \\
\times {}^{(D,\epsilon)}F_{2K+2}^{l_2 l_1} \left(\frac{\pi}{2}\right) {}^{(D,\epsilon)}F_{2K'+2}^{l_2' l_1} \left(\frac{\pi}{2}\right) N_{2K+2}^{l_2, l_1} N_{2K'+2}^{l_2', l_1} \\
\times 2^{-(l_2 + l_2')/2 - l_1} (1 + y)^{(l_2 + l_2')/2 - 1} (1 - y)^{l_1} \\
\times P_{(2K+2 - l_2 - l_1)/2}^{l_1 + \frac{1}{2}, l_2 + \frac{1}{2}}(y) P_{(2K'+2 - l_2' - l_1)/2}^{l_1 + \frac{1}{2}, l_2' + \frac{1}{2}}(y).
\end{aligned}
$$

(5.66)

Selection rule: In this case, $l = 2$, $l' = 2$, and $l'' = 2$. Hence, from Eq. (5.49), we see that the GSCs for coupling of two D states through tensor interaction vanish unless

$$
\max(K - K' - 1, K' - K - 1, 0) \le K'' \le (K + K' + 1).
$$

(5.67)

Once again, we have a finite number $(n_{KK'})$ of GSCs for fixed K and K' and the finite set of LIE, Eq. (5.66), can be solved by choosing $n_{KK'}$ different values of y in the interval $[-1, 1]$.

Sum rule: As in the earlier cases, we set $y = 1$ in Eq. (5.66) to get sum rules for $\epsilon = +$ and $\epsilon = -$ as

$$
\sum_{K''=\max(K-K'-1,K'-K-1,0)}^{K+K'+1} {}_D\langle K, +|K'', T|K', +\rangle_D
$$

$$
= -\frac{5}{2\pi}\sqrt{\frac{35}{2}}\, N^{(D,+)}_{2K+2}\, N^{(D,+)}_{2K'+2} \begin{pmatrix} 2\,2\,2 \\ 0\,0\,0 \end{pmatrix} \begin{bmatrix} 2\,2\,2 \\ 2\,0\,2 \end{bmatrix}
$$

$$
\times\, {}^{(D,+)}F^{2,0}_{2K+2}\left(\frac{\pi}{2}\right)\, {}^{(D,+)}F^{2,0}_{2K'+2}\left(\frac{\pi}{2}\right) (K+2)(K'+2)
$$

$$
\times \left[\frac{(K+1)(K+3)(K'+1)(K'+3)}{(2K+3)(2K+5)(2K'+3)(2K'+5)}\right]^{\frac{1}{2}}
$$

and

$$
\sum_{K''=\max(K-K'-1,K'-K-1,0)}^{K+K'+1} {}_D\langle K, -|K'', T|K', -\rangle_D = 0. \tag{5.68}
$$

5.4.6 Numerical Computation of GSCs

In the last five subsections, we have listed all the GSCs needed for the trinucleon with central plus tensor two-body interactions, together with a set of linear inhomogeneous equations for calculation of the GSCs and a sum rule for each case. Very fast numerical algorithms for solving the set of LIE are available. The advantages of the LIE method are threefold. *All the nonvanishing GSCs* for a particular coupling type, corresponding to given values of K and K', are obtained just by solving the set of LIE *only once*. In addition, this method is very fast and also very accurate. By contrast, a direct integration will involve a single, double, or triple sum over intermediate l values of the numerical integration of the product of three Jacobi polynomials of large orders (for large K and K'). Thus the integral has to be evaluated many times for a *single* GSC. Clearly, this process will be very slow and inaccurate for large K and K'. Alternately, a double sum (derived by Fabre [8] for the S-S′ coupling only) can be used for each (K, K', K'') combination. However, this double sum involves an alternating series, whose terms contain ratios of gamma functions of large arguments. This introduces large errors.

The LIE method has another advantage. It generates a sum rule, which provides an estimate of the error in computed GSCs. Numerical calculations [7] show that the percentage error for the sum rule is less than 10^{-8} % for coupling of S and S′ states through central forces. The error is about 10^{-6} % or less when an S or S′ state is coupled with the D state through tensor interactions. It is also in the same range for coupling of two D states through central or tensor forces. Details can be found in Ref. [7]. GSCs needed for the Reid soft-core (RSC) potential have been calculated by this method by Das and Bhattacharyya [9]. Thus very accurate calculations are possible using the LIE method for the GSCs, together with analytical results or accurate numerical quadratures for the potential multipoles.

5.5 Results of Numerical Calculations for ^{3}H and ^{3}He

As an example, we quote here some of the results reported by Ballot and Fabre [1]. They chose four semi-realistic nucleon–nucleon potentials proposed by Gogny, Pires, and de Tourreil (GPDT) [10] and by Sprung and de Tourreil (SSCA, SSCB, SSCC) [11]. All these potentials reproduce the two-nucleon data (properties of the deuteron and two-nucleon phase shifts) well, but differ slightly in the strengths and ranges of the central, tensor, LS, and L^2 terms. These potentials are called super-soft-core potentials, as each of them has a relatively soft short-range repulsion. Let us first look at the rate of convergence of the HH expansion. The sum over K in Eq. (5.36) was restricted to a maximum of K_{max}. Calculated energy (E), incremental energy (ΔE), and the matter radius (R_m) for different K_{max} values are presented in Table 5.1 for the GPDT potential, as an example. From this table, it is seen that both E and R_m converge fairly rapidly as K_{max} increases. Corresponding ΔE decreases rapidly to zero. We see that for $K_{max} = 12$, the energy converges to three significant digits. For better precision, one can go to larger values of K_{max}, but that would result in a much slower computation. As an alternative, one can derive a *convergence formula* for a specific type of potential [12] and use it together with the actually computed results for a few smaller K_{max} values to extrapolate the result for $K_{max} \to \infty$. The method of extrapolation will be discussed in detail in Sect. 1 of Chap. 6.

Table 5.1 Convergence of HH expansion for ^{3}H nucleus with GPDT potential

K_{max}	E	ΔE	R_m
2	−6.614		1.73
3	−7.509	−0.894	1.70
4	−7.977	−0.469	1.70
5	−8.248	−0.271	1.717
6	−8.395	−0.146	1.727
7	−8.468	−0.074	1.737
8	−8.515	−0.047	1.745
9	−8.543	−0.028	1.752
10	−8.559	−0.016	1.757
11	−8.569	−0.010	1.761
12	−8.575	−0.006	1.764

Energy and incremental energy (both in MeV) and matter radius (in fm) are presented for different K_{max} (corresponding to maximum grand orbital $L = 2K_{max}$)

Table 5.2 Results for ^{3}H nucleus with different two-body potentials by the HH expansion method

Potential	$-E(^3H)$ Calculated (MeV)	$-E(^3H)$ Extrapolated (MeV)	E_c (MeV)	$P(S)$	$P(S')$	$P(D)$	R_m
GPDT	8.58	8.58	0.66	94.3	0.97	4.72	1.77
SSCA	7.44	7.51	0.645	93.5	0.76	5.7	1.76
SSCB	7.34	7.41	0.65	93.8	0.81	5.4	1.78
SSCC	7.01	7.13	0.68	92.2	0.85	6.98	1.81
Experiment	8.48						1.7

Binding energy $[-E(^3H)]$ calculated with $K_{max} = 12$ and extrapolated, Coulomb energy (E_c), percentages of S, S', and D components of the wave function [$P(S)$, $P(S')$, and $P(D)$, respectively] and matter radius (R_m) for selected semi-realistic potentials. Energies are in MeV and radius is in fm

Table 5.2 lists the results of calculated binding energy $[-E(^3H)]$, both by direct calculation with $K_{max} = 12$ and the extrapolated value for the four selected potentials for ^{3}H nucleus. Fourth column presents the Coulomb energy. It is calculated as the difference of ground state energies of ^{3}He and ^{3}H nuclei. Fifth, sixth, and seventh columns present the percentage probabilities of S, S', and D states in the ground state of ^{3}H, denoted by $P(S)$, $P(S')$, and $P(D)$, respectively. The last column gives the calculated matter radius (R_m) of the triton. The last line presents the available experimental values.

The hyperspherical harmonics technique has been used in a variety of calculations for the trinucleon and other light nuclear systems, some of which are mentioned here. Coupled HH basis was used by Fang et al. to investigate photo effects on the isospin $\frac{3}{2}$ state of the trinucleon [13]. The realistic Reid soft-core (RSC) nucleon–nucleon potential was used by Bhattacharyya and Das to calculate trinucleon observables [14]. An efficient numerical algorithm for calculation of matrix element of the realistic RSC potential for the trinucleon ground state can be found in Ref. [15]. Triton asymptotic normalization constant was calculated by Ghosh and Das [16]. High precision was reported for realistic interactions by Kievsky et al. [17]. Accuracy of HHEM in the calculation of photodisintegration of triton was investigated by Barnea et al. [18]. Trinucleon energy levels were also calculated by Purcell et al. [19]. A high-precision variational calculation using a large HH basis was reported by Kievsky et al. [20]. Electromagnetic structure and reactions from chiral effective field theory were studied using trinucleon wave function calculated by the HHEM [20, 21]. Hyperspherical effective interaction for nonlocal two-body forces was calculated by Barnea et al. [22].

The HHEM has also been applied to the scattering problem involving the trinucleon [23]. Radiative captures of thermal neutrons in n-d and n-^{3}He were calculated by Girlanda et al. using wave functions obtained by HHEM with two-nucleon and three-nucleon realistic potentials [24].

5.6 Addition of Three-Nucleon Forces

In Table 5.2, we listed the results of four very simple model two-nucleon (2BF) potentials. Out of these the GPDT overbinds the trinucleon. For more realistic 2BF, both ^{3}H and ^{3}He are underbound by about 1.3 ± 0.3 MeV. More striking disagreement is in the calculated charge form factor (CFF), $F_{ch}(q^2)$. The first diffraction minimum of the calculated CFF occurs at $q^2 \approx 16$ fm^{-2} [1], while the experimental value is ≈ 11.8 fm^{-2}. The magnitude (F_{max}) of first maximum after this minimum has an experimental value of $\approx 6 \times 10^{-3}$, while the calculated value with 2BF is $\leqslant 1 \times 10^{-3}$. Thus, although the realistic 2BF reproduces all two-nucleon experimental data correctly, the same 2BF fails to reproduce trinucleon bound state properties. It immediately shows that the missing binding energy (BE) is due to three-nucleon force (3BF), which becomes active when three nucleons interact simultaneously. Hence, the net interaction is not just the sum of three pair-wise interactions. The 3BF is mediated by the exchange of two pions between the three nucleons: a pion emitted by the first nucleon is absorbed by the second nucleon, which is excited to the Δ resonance. The latter is subsequently de-excited with the emission of another pion, which is absorbed by the third nucleon. Due to the intermediate Δ resonance excitation, the Feynman diagram cannot be cut into two simple 2BF diagrams. Thus the 3BF is different from the sum of three pairs of 2BF. This process is referred to as the two-pion exchange (TPE) 3BF. This three-body force can easily be incorporated in the hyperspherical harmonics expansion technique. A large number of such calculations investigating the effects of 3BF on various properties of the trinucleon have been reported [25]. In the following we briefly discuss the method and the results obtained in the original simple calculation of Das et al. [26].

Since the S state has a probability of over 90 % for the ground state of the trinucleon, for a simple first calculation the effect of 3BF can be taken for the S state only. The effective TPE-3BF from the Fujita–Miyazawa force [27] acting on the S state of the trinucleon has the form [26]

$$V^{(3)}(\xi, \Omega_6) = \sum_{i,j,k=1,2,3 \text{ (cyclic)}} C_p (3 \cos^2 \theta_k - 1) U_{(2)}(x_i) U_{(2)}(x_j), \qquad (5.69)$$

where θ_k is the angle between the directions $\vec{x}_j$ and $\vec{x}_i$, where $\vec{x}_i = \vec{r}_j - \vec{r}_k$ (i, j, k cyclic). C_p is the coupling constant, whose value is in the range of $0.46 - 1.3$ MeV, according to the $\pi N \Delta$ coupling constant used in the calculation of TPE-3BF. $U_{(2)}(x)$ is given by

$$U_{(2)}(x) = \left[1 + \frac{3}{\mu x} + \frac{3}{(\mu x)^2}\right] \frac{e^{-\mu x}}{\mu x}, \qquad (5.70)$$

where μ is the inverse of the range of two-pion exchange 2BF, whose numerical value is 0.7 fm^{-1}, being proportional to the reciprocal of pion mass in theoretical units. The 3BF can be added with the standard 2BF in the Schrödinger equation. Since we are at present interested only in the effect of the 3BF on the S state of the trinucleon, we can expand the effective 3BF, Eq. (5.69), in the HH multipoles and add the multipole of the 3BF with the corresponding multipole of the 2BF in the calculation of the potential matrix. After this we proceed as before. The multipoles of the 3BF are given by [26]

$$
v^{(3)}_{2K''}(\xi) = c_p \frac{\pi^{\frac{3}{2}}}{128} \frac{(-1)^{K''}}{\langle 0,0|K'',C|K'',0\rangle} N^{(0)}_{2K''} \sum_{l=0,\text{even}}^{K''} \sqrt{2l+1} \,^{(0)}F^{ll}_{2K''}\left(\frac{\pi}{2}\right)
$$

$$
\times \sum_{K_1,K_2} (-1)^{K_1+K_2} U_{2K_1}(\xi) U_{2K_2}(\xi) \sum_\lambda (2\lambda+1) \begin{pmatrix} 2 & l & \lambda \\ 0 & 0 & 0 \end{pmatrix}^2
$$

$$
\times \left\langle \,^{(2)}\mathcal{P}^{2,0}_{2K_1+2}|\,^{(2)}\mathcal{P}^{l,l}_{2K''}|\,^{(2)}\mathcal{P}^{\lambda,l}_{2K_2+2}\right\rangle
$$

$$
\times \,^{(2)}\mathcal{P}^{2,0}_{2K_1+2}(0)\,^{(2)}\mathcal{P}^{\lambda,l}_{2K_2+2}(-2\pi/3), \tag{5.71}
$$

where $c_p = (m/\hbar^2)C_p$ and

$$
U_{2K}(r) = \frac{2^7}{15\pi}\left(K+\frac{5}{2}\right)\left(K+\frac{3}{2}\right)\int_0^1 U_{(2)}(rr')\,_2F_1\left(-K, K+4; \frac{7}{2}; r'^2\right)\sqrt{1-r'^2}\, r'^4 dr'. \tag{5.72}
$$

The final Schrödinger equation for the triton including both 2BF and 3BF is given by [see Eqs. (3.30) and (3.36)]

$$
\left[-\frac{\hbar^2}{m}\frac{d^2}{d\xi^2} + \frac{\hbar^2}{m}\frac{\mathcal{L}_K(\mathcal{L}_K+1)}{\xi^2} - E\right]u_K(\xi)
$$

$$
+3\sum_{K',K''}\left[\langle K,0|K'',C|K',0\rangle\{v^{(2)}_{2K''}(\xi) + v^{(3)}_{2K''}(\xi)\}\right]u_{K'}(\xi) = 0, \tag{5.73}
$$

where the potential multipole is written as a sum of two-body multipoles [renamed as $v^{(2)}_{2K''}(\xi)$] plus three-body multipoles $v^{(3)}_{2K''}(\xi)$. This Schrödinger equation is solved in the usual fashion, subject to appropriate boundary conditions to obtain the energy (E) and the hyper-partial waves [$u_K(\xi)$].

To study the enhancements of BE and CFF due to the inclusion of 3BF, the two-nucleon potential was chosen as Afnan–Tang S3 potential [28], which is a simple semi-realistic soft-core potential. Equation (5.69) shows that the 3BF is attractive for the equilateral configuration ($\theta_1 = \theta_2 = \theta_3 = \pi/3$) and repulsive for linear configuration ($\theta_1 = \theta_2 = 0$ and $\theta_3 = \pi$). Then Eqs. (5.70–5.72) show that the 3BF has an attractive singularity, which goes as ξ^{-6} at the origin, for the triangle configuration. Since a softcore 2BF is used, this singularity makes the Hamiltonian unbound from below. For a fully realistic 2BF, the very strong repulsion between nucleons at short separation prevents this catastrophy. As a softcore 2BF was used, a purely phenomenological cut-off parameter (x_0) was used to restrict the 3BF at short separations:

Table 5.3 Enhancement effects of Fujita–Miyazawa 3BF on the BE and CFF of the trinucleon

Description	C_p (MeV)	x_0 (fm)	BE (MeV)	$\|F_{ch}(q^2)\|$ Position of first zero (fm^{-2})	$F_{max} \times 10^3$	rms charge radius (fm)
^{3}H (2BF) (Calc.)			6.489	15.98	1.50	1.82
^{3}H (2BF+3BF) (calc.)	0.9	0.340	7.658	16.47	1.94	1.74
^{3}He (2BF) (calc.)			5.789	15.91	1.06	1.89
^{3}He (2BF+3BF) (calc.)	0.9	0.340	6.922	16.39	1.39	1.81
	0.46	0.277	6.485	15.54	1.58	1.84
^{3}H (expt.)			8.482			1.70±0.05
^{3}He (expt.)			7.718	11.8	∼ 6	1.84±0.03

Only zero node results are quoted from Ref. [26]

$$U_{(2)}(x) = \begin{cases} U_{(2)}(x_0), & x < x_0 \\ U_{(2)}(x), & x \geq x_0. \end{cases} \tag{5.74}$$

The effect of 3BF was studied treating x_0 as a parameter. For x_0 less than a critical value, there are nodes near the origin. Table 5.3 presents some of the enhancement effects due to the inclusion of 3BF. Only zero node cases are quoted from Ref. [26]. It can be seen that the addition of 3BF enhances the BE substantially toward the experimental value. However, CFF is not much improved. F_{max} increases, but the increase is not enough. Moreover, the position of the first diffraction minimum remains stubbornly too far to the right. Inclusion of 3BF brings the charge radii within the error bars of the experimental values. The reasons that the inclusion of 3BF fails to reproduce the experimental results completely are the following. The two-nucleon potential chosen in this simple calculation is not realistic. A realistic potential like the Reid soft-core (RSC) potential with a properly strong short-range repulsion (which will dispense with the arbitrary cut-off parameter x_0) should be used. Furthermore, the Fujita–Miyazawa form of two-pion exchange three-nucleon potential is not realistic, as it has an attractive singularity. This has been replaced by more 'tamed' three-nucleon potential derived incorporating nucleon form factors [25]. Later detailed calculations using more realistic 2BF plus 3BF reproduced the missing BE, but fell short of reproducing the experimental CFF [25]. The experimental charge form factor was explained by including the effects of meson exchange currents in the Feynman diagrams representing electron–trinucleon scattering processes [29].

References

1. Ballot, J.L., Fabre de la Ripelle, M.: Ann. Phys. (N.Y.) **127**, 62 (1980)
2. Das, T.K., Coelho, H.T., Torreao, J.R.A.: Phys. Rev. C **45**, 2640 (1992)
3. Rose, M.E.: Elementary Theory of Angular Momentum. Wiley, New York (1967)
4. Raynal, J., Revai, J.: Nuovo Cimento A **68**, 612 (1970)
5. Erens, G., Visschers, J.L., Van Wageningen, R.: Ann. Phys. (N.Y.) **67**, 461 (1971)
6. Das, T.K., De, T.B.: Pramana (J. Phys.) **28**, 645 (1987)
7. De, T.B., Das, T.K.: Phys. Rev. C **36**, 402 (1987)
8. Fabre de la Ripelle, M.: Ann. Phys. (N.Y.) **147**, 281 (1983)
9. Das, T.K., Bhattacharyya, S.: Pramana (J. Phys.) **40**, 189 (1993)
10. Gogny, D., Pires, P., de Tourreil, R.: Phys. Lett. B **32**, 591 (1970)
11. de Tourreil, R., Sprung, D.W.L.: Nucl. Phys. A **201**, 193 (1973)
12. Ballot, J.L., Beiner, M., Fabre de la Ripelle, M.: In: Calogero, F., Ciofi degli Atti, Y.C. (eds.) The Nuclear Many Body Problem, vol. 1. Bologna (1973); Fabre de la Ripelle, M. et al.: Lett. al Nuo. Cim. **1**, 584 (1971)
13. Fang, K.K., Levinger, J.S., Fabre de la Ripelle, M.: Phys. Rev. C **17**, 24 (1978)
14. Bhattacharyya, S., Das, T.K.: Pramana (J. Phys.) **44**, 183 (1995); Bhattacharyya, S., Das, T.K., Kanta, K.P., and Ghosh, A.K.: Phys. Rev. C **50**, 2228 (1994)
15. Bhattacharyya, S., Das, T.K.: J. Comput. Phys. (USA) **114**, 308 (1994)
16. Ghosh, A.K., Das, T.K.: Phys. Rev. C **42**, 1249 (1990)
17. Kievsky, A., et al.: Few-Body Syst. **22**, 1 (1997)
18. Barnea, N., et al.: Few-Body Syst. **39**, 1 (2006)
19. Purcell, J.E., et al.: Nucl. Phys. A **848**, 1 (2010)
20. Kievsky, A., et al.: J. Phys. G **35**, 063101 (2008); Marcucci, L.E., et al., Phys. Rev. C **83**, 014002 (2011)
21. Girlanda, L., et al.: J. Phys. Conf. Series **527**, 012022 (2014); Piarulli, M., et al., (*52 Int. Winter Meeting on Nuclear Physics, Bormio, 2014*) Proc. of Sci. (Bormio 2014) 040 (2014)
22. Barnea, N., Liedemann, W., Orlandini, G.: Phys. Rev. C **81**, 064001 (2010): Orlandini, G., Barnea, N., Leidemann, W.: J. Phys. Conf. Series **312**, 092049 (2011)
23. Viviani, M., et al.: arXiv:1109.3625 [nucl-th] (2011); Kovalchuk, V.I., Kozlovskii, I.V.: Phys. Part. Nucl. **43**, 294 (2012); Marcucci, L.E., et al.: Phys. Rev. C **80**, 034003 (2009); Kievsky, A., et al.: Few-Body Syst. **54**, 2395 (2013)
24. Girlanda, L., et al.: Phys. Rev. Lett. **105**, 232502 (2010)
25. Das, T.K., et al.:Lett. al Nuo. Cim. **33**, 1, (1982); Coelho, H.T. et al.: Phys. Letts. B **109**, 255 (1982); Das, T.K., et al.: Phys. Rev. C **26**, R754 (1982); Phys. Rev. C **26**, 697 (1982); Coelho, H.T., et al.: Phys. Rev. C **28**, 1812 (1983); Robilotta, M.R., et al.: Phys. Rev. C **31**, 646 (1985); De, T.B., Das, T.K.: Pramana J. Phys. **30**, 183 (1988); Ghosh, A.K., Das, T. K.: Pramana J. Phys. **36**, 313 (1991); Kievsky, A., et al.: Few-Body Syst. **45**, 115 (2009); Phys. Rev. C **81**, 044003 (2010); Viviani, M., et al.: Few-Body Syst. **54**, 885 (2013). Phys. Rev. Lett. **111**, 172302 (2013)
26. Das, T.K., Coelho, H.T., Fabre de la Ripelle, M.: Phys. Rev. C **26**, 2288 (1982)
27. Fujita, J., Miyazawa, H.: Prog. Theor. Phys. **17**, 360 (1957)
28. Afnan, I.R., Tang, Y.C.: Phys. Rev. **175**, 1337 (1968)
29. Kanta, K.P., Das, T.K.: Pramana J. Phys. **39**, L405 (1992); **43**, 313 (1994); Fizika B **5**, 4 (1996)

Chapter 6
Application to Coulomb Systems

Abstract Few-body Coulomb systems are discussed as examples of the hyperspherical harmonics technique. Nearly complete analytical calculation of the potential matrix element is possible for the Coulomb interaction. As a simple illustration, two-electron atoms are treated in details. There is no approximation except an upper cut-off of the HH basis, which is tested for convergence of binding energy (BE). An extrapolation formula can be obtained for the BE corresponding to the complete basis, from BE calculated with a few truncated basis functions. Hence very high precision is possible. General three-body Coulomb system with adiabatic approximation is also presented. Applications of these methods to physical systems are discussed.

Atoms and molecules are examples of Coulomb systems, in which the dominant interaction is the Coulomb interaction. As basic building blocks of all matter, they play a major role in physics and are of great interest. Coulomb interaction also exists in nuclear systems containing more than one charged nuclei, but it is of secondary importance compared to the dominant strong nuclear interactions. The majority of atoms and molecules contain a large number of particles, for which the hyperspherical harmonics expansion technique is not very practical. For such large Coulomb systems, several approximate many-body techniques have been developed, e.g., the semi-classical Thomas-Fermi (TF) model [1, 2], Hartree-Fock (HF) method [1, 3, 4], density functional theory (DFT) [5], etc. Both TF and HF models are based on mean-field approaches. TF model is simpler and uses statistical and semi-classical ideas. The HF theory provides the mean-field more elaborately in a self-consistent manner. On the other hand, DFT uses functionals of particle densities directly, avoiding a complete solution of the Schrödinger equation. Since atoms and molecules are fundamental building blocks of all stable matter, a particularly pertinent question concerns the stability of their bound states, either free or in the presence of electromagnetic fields. Rigorous quantum mechanical analysis of the stability of large Coulomb systems can be found in the lecture note series [6]. Spins of the constituent particles complicate the treatment as appropriate symmetrization has to be imposed. Condition of existence of the bound ground state of such a system is of special importance. Substantial amount of mathematically rigorous information concerning the stability has been obtained in the recent past [7–12].

© Springer India 2016

T.K. Das, *Hyperspherical Harmonics Expansion Techniques*,
Theoretical and Mathematical Physics, DOI 10.1007/978-81-322-2361-0_6

In this monograph, we are concerned with the application of hyperspherical harmonics to physical systems. As this technique is not practical for systems containing a large number of particles, its use is restricted to few-body Coulomb systems only. Other hyperspherical approaches, which are somewhat different from the technique presented in Chaps. 3 and 4, have also been adopted. A method using generalized Sturmian basis functions, can be found in Ref. [13]. Use of row-orthonormal hyperspherical coordinates for triatomic [14] and tetra-atomic [15] systems have also been proposed. As a simple example, we consider only three-body Coulomb systems in this chapter, for which ab initio calculation without any approximation is possible. Due to the simple analytic form of Coulomb interaction, a large part of calculation of potential matrix elements can be done analytically. This permits high degree of accuracy in numerical results. On the other hand, high precision spectroscopic data, especially for two-electron atoms, provide very accurate experimental results to compare with theoretical ones.

A three-body Coulomb system is a bound state of three charged particles—two of same charge and the third of the opposite charge—bound by the net Coulomb attraction. In such systems, all the particles can be identical (denoted by $A^+A^-A^+$), or two can be identical (denoted by $A^+B^-A^+$) or all different (denoted by $A^+B^-C^+$). Charge conjugated systems like $A^-B^+C^-$ will have the same properties as those of the $A^+B^-C^+$ system. An example of $A^-A^+A^-$ is the negatively charged positronium ion Ps^- (which is $e^-e^+e^-$). Spectroscopically very interesting class of two-electron atoms like the neutral He-atom, singly ionized lithium atom (Li^+), doubly ionized beryllium atom (Be^{++}), etc. are examples of the second class. Other examples of this class are: $e^-\mu^+e^-$ (negatively charged muonium atom), $d^+\mu^-d^+$, H^-–ion ($p^+e^-e^-$), hydrogen molecular ion (H_2^+, which is $p^+p^+e^-$), etc. Examples of $A^+B^-C^+$ systems are: HD^+ ($d^+p^+e^-$) and HT^+ ($t^+p^+e^-$), etc.

6.1 Two-Electron Atoms

In the following, we take the nucleus of two-electron atoms as infinitely heavy, so that it remains at rest. Even for the H^-, the nucleus to electron mass ratio, m_H/m_e, is about 1837. For heavier atoms, this ratio is even larger. Hence, this assumption is quite justified and it makes the calculation simpler. However, such an assumption is not essential and the nuclear motion can easily be incorporated using the Jacobi vectors introduced in Chap. 3 and the analytical treatment adopted here. For the Coulomb force, the matrix element can be analytically simplified to a great extent.

Thus, we take the nucleus fixed at the origin and the two electrons are at $\vec{r}_1$ and $\vec{r}_2$. We choose atomic units (a.u.), in which lengths are in units of Bohr radius (Bohr), $a_0 = \hbar^2/(m_e e_0^2)$, with $e_0^2 = e^2/(4\pi\epsilon_0)$. Its numerical value is $0.5291772083 \times 10^{-10}$ m [16]. Energy is expressed in atomic units (a.u.) of energy, e_0^2/a_0, whose numerical value is 27.2113834 eV [16]. In the atomic unit the Schrödinger equation for a two-electron atom with nuclear charge Ze is

$$\left[-\frac{1}{2}\nabla_1^2 - \frac{1}{2}\nabla_2^2 - \frac{Z}{r_1} - \frac{Z}{r_2} + \frac{1}{|\vec{r}_1 - \vec{r}_2|} - E \right]\Psi(\vec{r}_1, \vec{r}_2) = 0. \qquad (6.1)$$

We first discuss an exact calculation without using adiabatic approximation for the ground state of a two-electron atom in Sect. 6.1.1. Later in Sect. 6.2.1, we will discuss more general Coulomb systems using adiabatic approximation.

6.1.1 Exact Non-adiabatic Treatment of Two-Electron Atoms

In this section, we discuss exact calculation of two-electron atoms without adiabatic approximation. To make it simpler, we disregard the nuclear motion, assuming the nucleus to be infinitely heavy. This approximation can easily be relaxed by using Jacobi vectors instead of the position vectors of the electrons with respect to the nucleus fixed at the origin. To solve Eq. (6.1), we first introduce hyperspherical variables: a hyperradius (r) and a hyperangle (ϕ) are defined through

$$r_1 = r \sin\phi,$$
$$r_2 = r \cos\phi. \qquad (6.2)$$

The set of five hyperangles are constituted by $\Omega \equiv \{\phi, \vartheta_1, \varphi_1, \vartheta_2, \varphi_2\}$, where $\{\vartheta_i, \varphi_i\}$ are the polar angles of $\vec{r}_i$. Note the similarity in the definition of hyperspherical variables and the six-dimensional Laplacian with those in Chap. 3—only the Jacobi vectors $\vec{\xi}_1, \vec{\xi}_2$ have been replaced by $\vec{r}_1, \vec{r}_2$ in the present case. Hence, the HH basis will be the same functions of the presently defined hyperspherical variables. The advantage in the definition (6.2) is that actual particle separations are used. The position variables ($\vec{r}_1, \vec{r}_2$) and the corresponding orbital angular momenta ($\vec{l}_1, \vec{l}_2$) are the physical variables of the unperturbed Hamiltonian (without the inter-electronic repulsion). Moreover, symmetry of the wave function only under P_{12} is relevant, which for the present definition is ($\phi \leftrightarrow \pi/2 - \phi$). The hyperradius r remains invariant under P_{12}. Total orbital angular momentum $\vec{l}$ ($= \vec{l}_1 + \vec{l}_2$), its projection m, total spin $\vec{S}$ ($= \vec{s}_1 + \vec{s}_2$), and its projection M_S are good quantum numbers. The angular momentum coupled HH, $\mathcal{Y}_{Kl_1l_2lm}(\Omega)$, is given by Eqs. (3.33) and (3.34) [where $K = 2n + l_1 + l_2$ with n a nonnegative integer]. Expansion of Ψ in the HH basis is given by

$$\Psi_{lm}(\vec{r}_1, \vec{r}_2) = r^{-\frac{5}{2}} \sum_{Kl_1l_2} u_{Kl_1l_2lm}(r)\, \mathcal{Y}_{Kl_1l_2lm}(\Omega). \qquad (6.3)$$

From Eqs. (3.33) and (3.34), the effect of P_{12} is found to be

$$P_{12}\mathcal{Y}_{Kl_1l_2lm}(\Omega) = (-1)^{(n+l_1+l_2-l)}\mathcal{Y}_{Kl_2l_1lm}(\Omega). \qquad (6.4)$$

Note that this is different from the choice of hyperspherical variables using Jacobi coordinates in Chap. 3, where the effect of P_{12} was $(-1)^{l_1}$. Substitution of Eq. (6.3) in (6.1) and projection on a particular HH gives the set of coupled differential equations (CDE)

$$
\left[-\frac{1}{2}\frac{d^2}{dr^2} + \frac{1}{2}\frac{\mathcal{L}_K(\mathcal{L}_K+1)}{r^2} - E\right]u_{Kl_1l_2lm} + \frac{1}{r}\sum_{K'l_1'l_2'}\left\langle K\alpha\right|\left[-Z\left(\frac{1}{\sin\phi} + \frac{1}{\cos\phi}\right)\right.
$$

$$
\left. + \frac{1}{[\sin^2\phi + \cos^2\phi - 2\sin\phi\cos\phi\cos\theta_{12}]^{\frac{1}{2}}}\right]\left|K'\alpha'\right\rangle u_{K'l_1'l_2'lm}(r) = 0, \qquad (6.5)
$$

where $\alpha \equiv \{l_1l_2lm\}$, $\alpha' \equiv \{l_1'l_2'lm\}$, $\mathcal{L}_K = K + \frac{3}{2}$ and θ_{12} is the angle between $\vec{r}_1$ and $\vec{r}_2$. There is no approximation in Eq. (6.5), except for the assumption of an infinitely heavy nucleus (which can be removed as stated above) and the eventual truncation of the expansion basis for a computer calculation. Solution of Eq. (6.5) subject to appropriate boundary conditions at the origin and at infinity gives energy of a state with orbital angular momentum l and its projection m. Energy eigenvalues are independent of m due to rotational degeneracy. The solution with the lowest energy corresponds to the ground state and the higher ones to hyperradial excitations.

The Hamiltonian is spin independent, but E depends on S through the symmetry requirement. Spins of two electrons couple to S, which takes the values 0 and 1 for the singlet and triplet states, respectively. The singlet and triplet states are respectively antisymmetric and symmetric in spin space. Since the electrons are identical fermions, the singlet and triplet states are to be combined with symmetric and antisymmetric space wave functions respectively. For the ground state, $l = 0$. Hence, $l_1 = l_2$ and $K = 2n + 2l_1$ must be an even integer. For the spin singlet states, space wave function must be symmetric under P_{12} and Eq. (6.4) demands that n must be a nonnegative even integer. Hence l_1 can take even integral values $0, 2, \ldots, K/2$, if $K/2$ is even and odd integral values $1, 3, \ldots, K/2$, if $K/2$ is odd. If the expansion (6.3) is truncated at $K = K_{max}$ and all allowed l_1 values retained, then the number of basis functions (which is the same as the number of CDE) is

$$
N = \begin{cases} (K_{max}/4 + 1)^2 & \text{if } K_{max}/2 \text{ is even} \\ (K_{max}/2 + 1)(K_{max}/2 + 3)/4 & \text{if } K_{max}/2 \text{ is odd} \end{cases} \qquad (6.6)
$$

The effective potential of the Hamiltonian is given by

$$
V(r, \phi, \theta_{12}) = \frac{C(\phi, \theta_{12})}{r}
$$

with

$$
C(\phi, \theta_{12}) = -Z\left(\frac{1}{\sin\phi} + \frac{1}{\cos\phi}\right) + [\sin^2\phi + \cos^2\phi - 2\sin\phi\cos\phi\cos\theta_{12}]^{-\frac{1}{2}} \qquad (6.7)
$$

A plot of this potential surface as a function of ϕ and θ_{12} for a fixed value of $r = 1$ and $Z = 1$ is shown in Fig. 6.1. The effective potential in the hyperradial space has

Fig. 6.1 Plot of the potential surface at a fixed hyperradius ($r = 1$) of a two-electron atom with $Z = 1$ as function of ϕ and θ_{12} (in radian) along x- and y-axes, respectively

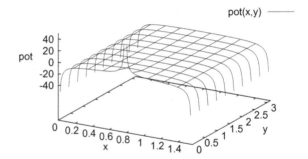

a r dependence of $\frac{1}{r}$, with a hyperangle dependent effective charge $C(\phi, \theta_{12})$. The potential surface becomes deeply attractive for $\phi \longrightarrow 0$ and $\phi \longrightarrow \pi/2$. These correspond to very strong Coulomb attraction, when either of the electrons approaches the nucleus. A steep repulsive spike appears at $\phi = \pi/4$ and $\theta_{12} = 0$, corresponding to the two electrons approaching each other. Finally, a saddle point appears at $\phi = \pi/4$ and $\theta_{12} = \pi$. At this point the potential increases as θ_{12} changes from π and the potential decreases as ϕ changes from $\pi/4$ in either direction. The saddle point is important for the quasi-stability of the doubly excited states.

The matrix elements in Eq. (6.5) can be simplified analytically [17] to expressions involving one-dimensional integrals to be evaluated numerically. The matrix element of electron-nucleus attraction for the $l = m = 0$ states is given by

$$
\left\langle Kl_1l_100 \left| \left(\frac{1}{\sin \phi} + \frac{1}{\cos \phi} \right) \right| K'l_1'l_1'00 \right\rangle
$$

$$
\equiv \delta_{l_1 l_1'} A^{l_1}_{KK'}
$$

$$
= \delta_{l_1 l_1'} \int_0^{\frac{\pi}{2}} {}^{(2)}\mathcal{P}^{l_1 l_1}_K(\phi) \, {}^{(2)}\mathcal{P}^{l_1 l_1}_{K'}(\phi) \sin \phi \cos \phi (\sin \phi + \cos \phi) d\phi. \tag{6.8}
$$

The inter-electron repulsion term can be simplified by using the generating function of Legendre polynomials, addition theorem of spherical harmonics [18], and properties of Clebsch–Gordan coefficients [19]. The result is

$$
\left\langle Kl_1l_100 \left| \frac{1}{[\sin^2 \phi + \cos^2 \phi - 2 \sin \phi \cos \phi \cos \theta_{12}]^{\frac{1}{2}}} \right| K'l_1'l_1'00 \right\rangle
$$

$$
\equiv B^{l_1 l_1'}_{KK'}
$$

$$
= (-1)^{l_1+l_1'} \sqrt{(2l_1+1)(2l_1'+1)} \sum_{n=|l_1-l_1'|}^{l_1+l_1'} \begin{pmatrix} l_1 & n & l_1' \\ 0 & 0 & 0 \end{pmatrix}^2
$$

$$
\times \int_0^{\frac{\pi}{4}} {}^{(2)}\mathcal{P}^{l_1 l_1}_K(\phi) \, {}^{(2)}\mathcal{P}^{l_1' l_1'}_{K'}(\phi) (\tan \phi)^n \cos \phi \sin^2 \phi d\phi. \tag{6.9}
$$

These matrix elements can be computed by suitable quadratures. In terms of these, the CDE takes the form

$$
\left[-\frac{1}{2}\frac{d^2}{dr^2} + \frac{1}{2}\frac{\mathcal{L}_K(\mathcal{L}_K+1)}{r^2} - E \right] u_{Kl_1 l_1' 00}(r)
$$
$$
+\frac{1}{r}\sum_{K' l_1'} \left[-Z\delta_{l_1 l_1'} A^{l_1}_{KK'} + 2B^{l_1 l_1'}_{KK'} \right] u_{K' l_1' l_1' 00}(r) = 0, \qquad (6.10)
$$

which can be solved by an accurate numerical algorithm, like the renormalized Numerov (RM) method [20], which will be discussed in Sect. 10.2.1 of Chap. 10.

6.1.2 Convergence of HH Expansion: Extrapolation and Accuracy

One can see from Eq. (6.6) that the number N of CDE increases rapidly with K_{max}. Since the computation time increases as N^3, this method will not be useful, unless the expansion converges fast. The rate of convergence for a system of bosons interacting through a single Gaussian potential has been studied [21]. In order to study the rate of convergence of the HH expansion in two-electron atoms, the method was applied to the ground state (1^1S state, corresponding to $l = 0$, $S = 0$) of H$^-$, He, Li$^+$, Be^{2+} and the first excited 2^1S state of He in Ref. [17]. Table 6.1 shows the calculated BE ($= B_{K_{max}} = -E$) for various K_{max} values. One notices that, although a clear convergence trend is discernible with increasing K_{max}, still BE does not reach the desirable precision for $K_{max} = 20$. The rate of convergence is much slower for the long-range Coulomb force than the short-range nuclear force as seen earlier. For larger K_{max}, the computation time increases rapidly. Moreover, both numerical errors and instabilities due to such errors increase with large N. Thus calculations for larger K_{max} become increasingly difficult. But BE for larger K_{max} can be extrapolated from the numerically calculated BE for smaller K_{max} values, using a convergence relation. Such a convergence formula for the Coulomb potential can be derived using

Table 6.1 Calculated BE (in a.u.) of some two-electron systems for various K_{max} values

K_{max}	H$^-$	He	Li$^+$	Be^{2+}	2^1S state of He
0	0.385397	2.500017	6.459134	12.262765	1.275519
4	0.480799	2.784369	7.039221	13.248454	1.599267
8	0.502585	2.850214	7.175991	13.482733	1.771541
12	0.512577	2.876006	7.227336	13.568884	1.878540
16	0.517726	2.887540	7.249755	13.606178	1.947698
20	0.520737	2.893580	7.261233	13.625078	1.994575

Schneider's theorem [22] on convergence of HH expansion [17], as

$$(K_{\max} + x)^4 \Delta B_{K_{\max}} = D, \tag{6.11}$$

where x and D are constants and

$$\Delta B_{K_{\max}} = B_{K_{\max}+4} - B_{K_{\max}}. \tag{6.12}$$

Relation (6.11) has been tested by actual calculations for smaller $K_{\max}$ values and then extrapolated for a large $K_{\max}$ value, consistent with computational error [17]. Table 6.2 presents the converged extrapolated binding energies thus obtained, together with other standard calculations, including very elaborate and precise calculations by Pekeris [24, 26].

In the last row of Table 6.2, we quote the result of a similar calculation for the positronium ion Ps$^-$ from Ref. [27]. One can see that the extrapolated HHEM calculations are almost as precise as those done in Refs. [24, 26, 28, 29], although the former is much simpler and faster. We can also observe that the relative accuracy of the HHEM calculations are better for the heavier atoms. This is due to the disregard of the nuclear motion in the extrapolated HHEM calculations, the assumption being more justified as the nucleus becomes heavier. An essentially exact many-body calculation for muonic molecular ions and other exotic Coulombic system was reported by Chakrabarti and Das [30]. This method has also been applied to exotic two-muon three-body Coulomb systems by Frolov and Wardlaw [31] and by Khan [32].

The HHEM was used together with the techniques of supersymmetric quantum mechanics [33] by Das and Chakrabarti to calculate binding energy (BE) of excited

Table 6.2 Comparison of calculated binding energies (in a.u.) of two-electron atoms by extrapolated HHEM with other theoretical calculations

Atom	State	Calculated BE (a.u.)		Expt.
		Extrapolated HHEM	Other calculation	
H$^-$	1^1S	0.52668 [17]	0.52621 [23] 0.52775 [24]	0.52777
He	1^1S	2.90368 [17]	2.903 [25] 2.90271 [23] 2.90372 [24]	2.9038
Li$^+$	1^1S	7.28007 [17]	7.27832 [23] 7.28008 [24]	7.2804
Be^{2+}	1^1S	13.65600 [17]	13.65319 [23] 13.65600 [24]	13.6572
He	2^1S	2.13895 [17]	1.998 [25] 2.14597 [26]	2.14606
Ps$^-$	1^1S	0.2616689 [27]	0.2620047 [28] 0.2620049 [29]	

states very accurately. The convergence of the HH expansion for the excited state is very slow, as the wave function gets more extended with excitation. For a given excited state the supersymmetric partner potential is calculated, whose ground state has the same energy as the chosen excited state. Calculation of the BE of the ground state in the partner potential converges very fast, giving an accurate BE for the excited state. The method was applied to the $^1S^e$ state of the He atom [34].

6.2 General Three-Body Coulomb Bound Systems

For a general Coulomb system $A^+B^+C^-$ with arbitrary masses m_A, m_B, m_C and charges Z_A, Z_B, Z_C (two of same sign and the third of the opposite sign) respectively, the Schrödinger equation is

$$\left[-\frac{1}{2m_A}\nabla_A^2 - \frac{1}{2m_B}\nabla_B^2 - \frac{1}{2m_C}\nabla_C^2 + \frac{Z_A Z_B}{r_{AB}} + \frac{Z_A Z_C}{r_{AC}} + \frac{Z_B Z_C}{r_{BC}} - E_{\text{Tot}} \right]$$
$$\times \psi(\vec{r}_A, \vec{r}_B, \vec{r}_C) = 0. \tag{6.13}$$

Here r_{ij} is the distance between particles i and j and E_{Tot} is the total energy of the system. We follow the treatment of the review article by C.D. Lin [35], in which the Jacobi vectors are chosen as

$$\vec{\rho}_1 = \vec{r}_{AB}$$
$$\vec{\rho}_2 = \vec{r}_{AB,C}, \tag{6.14}$$

where $\vec{r}_{AB}$ is the vector from particles A to B and $\vec{r}_{AB,C}$ is the vector from the center of mass of particle-pair A, B to particle C. In this coordinate system, the Schrödinger equation for the relative motion becomes

$$\left[-\frac{1}{2\mu_1}\nabla_{\vec{\rho}_1}^2 - \frac{1}{2\mu_2}\nabla_{\vec{\rho}_2}^2 + \frac{Z_A Z_B}{r_{AB}} + \frac{Z_A Z_C}{r_{AC}} + \frac{Z_B Z_C}{r_{BC}} - E \right]\Psi(\vec{\rho}_1, \vec{\rho}_2) = 0, \tag{6.15}$$

where E is the energy of the relative motion and

$$\mu_1 = \mu_{AB} = \frac{m_A m_B}{m_A + m_B}$$
$$\mu_2 = \mu_{AB,C} = \frac{(m_A + m_B)m_C}{m_A + m_B + m_C}. \tag{6.16}$$

In the above, no particle is assumed to be at rest. However, the choice of Jacobi vectors is different from that of Chap. 3, where the coefficients of the two kinetic energy terms were chosen to be the same. One can get the coordinate system similar to the one in Chap. 3 by using $\vec{\xi}_i = \sqrt{(\mu_i/\mu)}\, \vec{\rho}_i\,(i = 1, 2)$

$$\left[-\frac{1}{2\mu}(\nabla_{\vec{\xi}_1}^2 + \nabla_{\vec{\xi}_2}^2) + \frac{C(\Omega_6)}{\xi} - E\right]\Psi(\xi, \Omega_6) = 0, \tag{6.17}$$

where $\Omega_6 \equiv \{\phi, \vartheta_1, \varphi_1, \vartheta_2, \varphi_2\}$, (ϑ_i, φ_i) being the polar angles of $\vec{\xi}_i$. The effective charge $C(\Omega_6)$ is obtained from the Coulomb terms of Eq. (6.15). One can then solve Eq. (6.17) exactly, following the standard hyperspherical harmonics expansion technique of Chap. 3, with the matrix elements of the effective Coulomb potential calculated in the manner of Sect. 6.1.1. However, for a simpler calculation the adiabatic approximation is widely used. This approximation will be discussed in detail in Sect. 10.2.3 of Chap. 10. Here we discuss briefly its application to Coulomb systems in the next subsection.

6.2.1 Adiabatic Approximation in Coulomb Systems

The adiabatic approximation is similar to the Born–Oppenheimer approximation (BOA) and will be discussed in detail in Chap. 10, Sect. 10.2. It assumes that the hyperradial motion is slow compared to the hyperangular motion and separates the latter adiabatically for a fixed value of ξ [35]. The wave function is expanded in the set of adiabatic wave functions $\{\Phi_\kappa(\xi, \Omega_6)\}$

$$\Psi(\xi, \Omega_6) = \sum_\kappa \xi^{-\frac{5}{2}} u_\kappa(\xi)\Phi_\kappa(\xi, \Omega_6). \tag{6.18}$$

The adiabatic wave function is obtained by solving the Schrödinger equation at a fixed value of ξ [taking $\mu = 1$ in Eq. (6.17)]

$$\left[\frac{\mathcal{L}^2}{\xi^2} + \frac{2C(\Omega_6)}{\xi}\right]\Phi_\kappa(\xi, \Omega_6) = \omega_\kappa(\xi)\Phi_\kappa(\xi, \Omega_6), \tag{6.19}$$

where $\mathcal{L}^2$ is given by Eq. (3.8). Substitution of Eqs. (6.18) and (6.19) in (6.17) gives a set of CDE for $u_\kappa(\xi)$

$$\left[-\frac{d^2}{d\xi^2} + \omega_\kappa(\xi) - 2E\right]u_\kappa(\xi) = \sum_{\kappa'}\left[2\langle\Phi_\kappa|\frac{d\Phi_{\kappa'}}{d\xi}\rangle\frac{d}{d\xi} + \langle\Phi_\kappa|\frac{d^2\Phi_{\kappa'}}{d\xi^2}\rangle\right]u_{\kappa'}(\xi). \tag{6.20}$$

The index κ in Eq. (6.19) characterizes different eigenvalues [called *channel potentials* $\omega_\kappa(\xi)$] and it is used to label different *channels*. The global size of the system is given by the hyperradial wave function $u_\kappa(\xi)$, while internal motion including overall rotation of the system is governed by the channel function $\Phi_\kappa(\xi, \Omega_6)$. Hyperradial excitations for the same channel potential in Eq. (6.20) correspond to the *breathing modes*. An important aspect of the solution of Eq. (6.19) is to identify the different modes of internal motion in terms of quantum numbers used to represent the channel index κ.

Exact solutions of Eqs. (6.19) and (6.20) can be obtained by the RM method (see Sect. 10.2.1 of Chap. 10). As we discussed already, for the two-electron atoms, $\vec{l}$ and $\vec{S}$ are good quantum numbers. This is also true for other three-body systems in which the particles interact only through Coulomb (in general spin-independent central) forces. In this case, it is particularly convenient to expand the channel function in products of spherical harmonics coupled to a total orbital angular momentum (l) for a particular spin angular momentum (S), in the following manner

$$\Phi_\kappa^{lmSM_S}(\xi, \Omega_6) = \hat{O}_{\text{Sym}} \sum_{l_1 l_2} f_{l_1 l_2}^{lS}(\xi, \phi)[Y_{l_1 m_1}(\vartheta_1, \varphi_1) Y_{l_2 m_2}(\vartheta_2, \varphi_2)]_{lm}, \quad (6.21)$$

where $\hat{O}_{\text{Sym}}$ represents the symmetrization $(\hat{S})$ or antisymmetrization $(\hat{A})$ operator, which selects appropriate values of (l_1, l_2) for specific values of l and S. The summation is over pairs of selected (l_1, l_2) values, such that the spatial wave function has the symmetry conjugate to that of the spin wave function, in order that the total wave function has the required symmetry. For example, for a two-electron atom, the space wave function should be symmetric or antisymmetric spin wave function. Note that the spin wave function does not appear explicitly in Eq. (6.21). For small ξ, the function $f(\xi, \phi)$ spreads over the entire interval $[0, \pi/2]$ of ϕ and it approaches the hyperspherical harmonics in the limit $\xi \to 0$. On the other hand, for large ξ, it is localized either in $\phi \to 0$ or $\phi \to \pi/2$, corresponding to either the first or the second electron being near the nucleus, with $f(\xi, \phi)$ approaching the hydrogenic wave function.

The set of Eqs. (6.19) and (6.20) has been solved by several methods:

1. Macek [36] solved these equations for the doubly excited states of the Helium atom by direct numerical integration. However, this method encounters numerical instabilities when eigenvalues are nearly degenerate.
2. Lin [37] diagonalized the effective Hamiltonian in the HH basis. The convergence of the expansion becomes slow at large values of ξ.
3. Lin [38] also solved the adiabatic equation by diagonalization, using analytic channel functions (hydrogenic basis functions for large ξ, generalized to smaller ξ together with hyperspherical harmonics) to obtain channel potential curves. The method is quite accurate, stable, and economical.
4. Tang et al. [39] used the generalized Numerov method, using a three term recurrence relation (see Sect. 10.2.1 of Chap. 10) and resulting in an eigenvalue equation.

$$AX = \omega_\kappa(\xi)BX, \quad (6.22)$$

where A and B are tridiagonal, nonsymmetric matrices with fixed (l_1, l_2). The channel potential is found by searching for zeros of the secular equation

$$\det |A - \omega_\kappa(\xi) B| = 0. \tag{6.23}$$

The method becomes quite fast and accurate, since $\omega_\kappa(\xi)$ are obtained by solving Eq. (6.23) iteratively from its knowledge at a smaller ξ. The accuracy goes as the sixth power of the step size and can be greatly enhanced with smaller step sizes.

Channel potential curves and details of the calculations for a number of Coulomb systems can be found in an excellent review article by Lin [35].

References

1. Bransden, B.H., Joachain, C.J.: Physics of Atoms and Molecules. Longman, New York (1983)
2. Lieb, E.H., Simon, B.: Adv. Math. **23**, 22 (1977)
3. McWeeny, R.: Methods of Molecular Quantum Mechanics. Academic Press, London (1992)
4. Bach, V.: Approximative theories for large Coulomb systems. In: Rauch, J., Simon, B. (eds.) Quasiclassical Methods. Springer, New York (1997); Heinzi C., et al.: Phys. Rev. A **76**, 052104 (2007)
5. Parr, R.G., Yang, W.: Density-Functional Theory of Atoms and Molecules. Oxford University Press, Oxford (1989); Jones, R.O., Gunnarsson, O.: Rev. Mod. Phys. **61**, 689 (1989); Argaman N., Makov, G.: Am. J. Phys. **68**, 69 (2000); Frank, R.L., Lieb, E.H., Seiringer, R., Siedentop, H.: Phys. Rev. A **76**, 052517 (2007)
6. Siedentop, H. (ed.): Complex Quantum Systems: Analysis of Large Coulomb Systems. Lecture Notes Series, vol. 24, Institute of Mathematical Sciences, National University of Singapore. World Scientific Publishing, Singapore (2013)
7. Lieb, E.H.: Many particle Coulomb systems, presented at the 1976 session on statistical mechanics of the International Mathematical Summer Center (C.I.M.E.). Bressanone, Italy 21–27 June 1976
8. Lieb, E.H., Loss, M.: Adv. Theor. Math. Phys. **7**, 667 (2003). arXiv:math-ph/0408001v1 31 July 2004
9. Benguria, R.D., Loss, M., Siedentop, H.: J. Math. Phys. **49**, 012302 (2008); Benguria, R.D., Frank, R.L., Loss, M.: Math. Res. Lett. **15**, 613 (2008)
10. Loss, M., Weidl, T.: J. Eur. Math. Soc. **11**, 1365 (2009); Frank, R.L., Loss, M.: J. Math. Pure Appl. **97**, 39 (2012)
11. Solovej, J.P.: In: Francoise, J.-P., Naber, G.L., Oxford, T.S.T. (eds.) Encyclopedia of Mathematical Physics, vol. 5, pp. 8–14. Elsevier (2006). Seiringer, R., Solovej, J.P.: Bull. AMS, **50**, 169 (2013)
12. Bellazzini, J., Frank, R.L., Lieb, E.H., Seilinger, R.: Rev. Math. Phys. **26**, 1350021 (2014)
13. Avery, J.S.: J. Comput. Appl. Math. **233**, 1366 (2010); Avery, J.S.: Hyperspherical Harmonics and Generalized Sturmians, Springer (2002); Avery, J., Avery, J.: Generalised Sturmians and Atomic Spectra, World Scientific (2007); Mitnik D.M., et al.: Comp. Phys. Comm. **182**, 1145 (2011)
14. Wang, D., Kuppermann, A.: J. Phys. Chem. A **113**, 15384 (2009)
15. Aquilanti, V., et al.: Int. J. Quant. Chem. **89**, 277 (2002); Aquilanti, V., Lombardi, A., Littlejohn, R.G.: Theo. Chem. Acc. **111**, 400 (2004); Kuppermann, A.: J. Phys. Chem. **108**, 8894 (2004); Lepetit, B., Wang, D., Kuppermann, A.: J. Chem. Phys. **125**, 133505 (2006); Kuppermann, A.: Phys. Chem. Chem. Phys. **13**, 8259 (2011)

16. Pethick, C.J., Smith, H.: Bose-Einstein Condensation in Dilute Gases. Cambridge University Press, Cambridge (2002)
17. Das, T.K., Chattopadhyay, R., Mukherjee, P.K.: Phys. Rev. A **50**, 3521 (1994)
18. Arfken, G.: Mathematical Methods for Physicists. Academic Press, New York (1966)
19. Roy, R.R., Nigam, B.P.: Nuclear Physics: Thory and Experiment. John Wiley and Sons Inc., New York (1967)
20. Johnson, B.R.: J. Chem. Phys. **69**, 4678 (1978)
21. Timofeyuk, N.K.: Phys. Rev. A **86**, 032507 (2012)
22. Schneider, T.R.: Phys. Lett. B **30**, 439 (1972)
23. Zhang, R., Deng, C.: Phys. Rev. A **47**, 71 (1993)
24. Pekeris, C.L.: Phys. Rev. **112**, 1649 (1958)
25. Ballot, J.L., Navarro, J.: J. Phys. B **8**, 172 (1975)
26. Pekeris, C.L.: Phys. Rev. **126**, 1470 (1962)
27. Chattopadhyay, R., Das, T.K., Mukherjee, P.K.: Phys. Scr. **54**, 601 (1996)
28. Krivac, R., Haftel, M.I., Mandelzwig, V.B.: Phys. Rev. A **47**, 911 (1993)
29. Haftel, M.I., Mandelzwig, V.B.: Phys. Rev. A **39**, 2813 (1989)
30. Chakrabarti, B., Das, T.K.: Molecular Phys. **107**, 1817 (2009)
31. Frolov, A.M., Wardlaw, D.M.: Eur. Phys. Jour. D **63**, 339 (2011)
32. Md. A. Khan, Few-Body Syst. **52**, 53 (2012); Eur. Phys. Jour. D **66**, 83 (2012); Few-Body Syst. doi:10.1007/s00601-014-0881-8 (published 04 April 2014); Int. J. Mod. Phys. E **23**, 1450055 (2014); arXiv:1406.4953v4 [atom-ph] 10 Jul 2014; arXiv:1412.7501v1 [atom-ph] 23 Dec 2014
33. Cooper, F., Khare, A., Sukhatme, U.: Phys. Rep. **251**, 267 (1995)
34. Das, T.K., Chakrabarti, B.: Phys. Rev. E **62**, 4347 (2000)
35. Lin, C.D.: Phys. Rep. **257**, 1 (1995)
36. Macek, J.: J. Phys. B **1**, 831 (1968)
37. Lin, C.D.: Phys. Rev. A **10**, 1986 (1974)
38. Lin, C.D.: Phys. Rev. A **23**, 1585 (1981)
39. Tang, J.Z., Watanabe, S., Matsuzawa, M.: Phys. Rev. A **46**, 2437 (1992)

Chapter 7
Potential Harmonics

Abstract Potential harmonics (PH) basis is the subset of HH, which is sufficient for expansion of two-body potential. It includes only two-body correlations. PH is appropriate for a sufficiently dilute system, in which only two-body correlations are relevant. Closed analytic expression for the PH is derived. Potential multipoles for this basis and overlap of PHs of different pairs are obtained. Use of symmetrized and unsymmetrized PH bases have also been discussed.

In earlier chapters, we discussed the expansion of the wave function in the complete set of hyperspherical harmonics (HH). We saw that the exact treatment becomes unmanageable for systems containing more than three particles. Even for the three-body system the degeneracy of the HH basis corresponding to a particular order (which is the grand orbital quantum number L) increases very rapidly with the order. For this reason the expansion basis has to be truncated according to several alternative schemes (Sect. 4.5 of Chap. 4). One of the truncation schemes is to use the potential harmonics (PH) basis. This is a subset of the HH basis which is complete for the expansion of the two-body potential $V(\vec{r}_{ij})$, justifying its name. Clearly, it depends on the label (ij) and the separation vector $\vec{r}_{ij}$ of the interacting pair. It is independent of the position vectors of all noninteracting particles. In order to express the full wave function (which depends on all particle coordinates) in the PH basis appropriate for (ij)-pair, we have to decompose the former in pair-wise Faddeev components

$$\psi(\vec{r}_1, \ldots, \vec{r}_A) = \sum_{i,j>i}^{A} \phi_{ij}(\vec{r}_1, \ldots, \vec{r}_A). \tag{7.1}$$

In general, the Faddeev component $\phi_{ij}(\vec{r}_1, \ldots, \vec{r}_A)$ corresponds to (ij)-interacting pair, but depends on position vectors of *all particles* in the system, and thereby taking care of all correlations. If we expand $\phi_{ij}(\vec{r}_1, \ldots, \vec{r}_A)$ in PH corresponding to (ij)-pair, then only its dependence on $\vec{r}_{ij}$ is retained, which corresponds to *only two-body correlation* and dependence on the relative configurations of three, four, ... particle subsystems in $\phi_{ij}(\vec{r}_1, \ldots, \vec{r}_A)$ is disregarded. Hence higher than two-body correlations are disregarded. Since $\psi(\vec{r}_1, \ldots, \vec{r}_A)$ is written as a sum of Faddeev components of *all interacting pairs*, all two-body correlations are properly accounted

© Springer India 2016
T.K. Das, *Hyperspherical Harmonics Expansion Techniques*,
Theoretical and Mathematical Physics, DOI 10.1007/978-81-322-2361-0_7

for. Note that the wave function ψ is uncorrelated if it is a product of single particle wave functions; it has two-body correlations if it depends on all pair-separations; it has three-body correlations if it depends on relative configuration of all three-particle sub-systems, and so on. To accommodate all two-body correlations in ψ, we take the Faddeev component $\phi_{ij}(\vec{r}_1, \ldots, \vec{r}_A)$ to be a function of $\vec{r}_{ij}$ and r (hyperradius) only, but independent of position vectors of all noninteracting particles. This means that all the inert degrees of freedom are frozen and the number of active variables reduces to *four only*, for each interacting-pair partition, corresponding to $\vec{r}_{ij}$ and r. This results in a great simplification for the treatment of the many-body system. The disregard of higher than two-body correlations is the price one pays for this simplification. When is this justified? Unless the system is very dense, the probability of finding more than two particles in a small volume will be very small. Therefore, the contribution of three- and higher-body correlations in ψ can be expected to be negligible and the use of PH basis is likely to be a good approximation. In very light and halo nuclei, where the density is quite low and average internucleon separation is large, this approximation can be expected to be fairly good. On the other hand, heavier nuclei have much shorter internucleon separations and such an approximation will not be good. For the Bose–Einstein condensate (see Chap. 8), which is a system containing several thousand to several million of bosonic atoms in a gas of extremely low density (number density $\sim 10^{15}$ cm^{-3}), the average interatomic separation is very large compared to the range of interatomic interaction and this approximation is very good.

7.1 Potential Harmonics

For simplicity, we again consider a system of A identical particles, each of mass m. The Jacobi vectors can be defined as in Eq. (4.1). However for our convenience, we rename the Jacobi vector $\vec{\xi}_j$ as $\vec{\xi}_{N-j+1}$ of Eq. (4.1), i.e.

$$\vec{\xi}_j = \left[\frac{2(N-j+1)}{N-j+2}\right]^{1/2}\left(\vec{r}_{N-j+2} - \frac{1}{N-j+1}\sum_{i=1}^{N-j+1}\vec{r}_i\right) \qquad (j = 1, \ldots, A-1),$$

$$(7.2)$$

where $N = A - 1$. Since Eq. (7.2) only renames the Jacobi vectors, all the relations of Sect. 4.1 of Chap. 4 will be valid. In particular, hyperradius and the hyperangles are defined through Eqs. (4.6) and (4.7), respectively. The HH is given by Eqs. (4.14) and (4.15). As already explained, we have to consider the interacting pair [say, the (ij)-pair] of particles, as a special active pair for the associated PH. Consequently, we again rename the particles i and j as 1 and 2 respectively, such that

$$\vec{\xi}_N = \vec{r}_{ij} = \vec{r}_j - \vec{r}_i.$$

$$(7.3)$$

We can do this, since all particles are identical. In doing this we can use the same equations for each interacting pair. However, we have to remember that the hyperangles depend on the interacting-pair label (ij). Hence, we denote the hyperangles by $\Omega_N^{(ij)}$.

The general HH is given by Eqs. (4.14) and (4.15). We wish to construct that subset of HH, which is sufficient for expansion of $V(\vec{r}_{ij})$, i.e. $V(\vec{\xi}_N)$. This potential in independent of all the Jacobi vectors $\vec{\xi}_1, \vec{\xi}_2, \ldots, \vec{\xi}_{N-1}$. Hence the desired subset, which is complete for the expansion of $V(\vec{r}_{ij})$ will not contain the Jacobi vectors $\vec{\xi}_1, \ldots, \vec{\xi}_{N-1}$. We can obtain such basis functions from the general HH by choosing the quantum numbers [see Eqs. (4.14)–(4.16)]

$$
\begin{aligned}
l_1 &= l_2 = \cdots = l_{N-1} = 0, & l_N &= l \\
m_1 &= m_2 = \cdots = m_{N-1} = 0, & m_N &= m \\
n_2 &= n_3 = \cdots = n_{N-1} = 0, & n_N &= K \\
L_1 &= L_2 = \cdots = L_{N-1} = 0, & L_N &= L = 2K + l.
\end{aligned} \tag{7.4}
$$

Thus, a member of this subset is given by

$$
\mathcal{P}_{2K+l}^{l,m}(\Omega_N^{(ij)}) \equiv \mathcal{Y}_{[L]}(\Omega_N^{(ij)}) = Y_{l,m}(\vartheta_N, \varphi_N) \,^{(N)}\mathcal{P}_{2K+l}^{l,0}(\phi_N) \mathcal{Y}_{[0]}(3N-3), \tag{7.5}
$$

where $\mathcal{Y}_{[0]}(3N-3)$ is the HH of order zero in $(3N-3)$-dimensional space, which it is a constant and its value is obtained from the normalization condition as

$$
\mathcal{Y}_{[0]}(3N-3) = \left[\frac{\Gamma((3N-6)/2)}{2\pi^{(3N-6)/2}} \right]^{\frac{1}{2}} \tag{7.6}
$$

The subset $\{\mathcal{P}_{2K+l}^{l,m}(\Omega_N^{(ij)})\}$ is called the *potential harmonics* (PH) *basis* corresponding to the (ij)-interacting pair. Note that it depends on the particular interacting pair (ij), as the set of hyperangles is denoted by $\Omega_N^{(ij)}$, identifying the pair (ij). Consequently, the angles ϑ_N, φ_N and ϕ_N of Eq. (7.5) should carry this identification, which we suppress for brevity. The potential harmonic $\mathcal{P}_{2K+l}^{l,m}(\Omega_N^{(ij)})$ of order $2K + l$ is a spherical tensor of rank l in 3-dimensional space and involves three quantum numbers l, m and the grand orbital $2K + l$, corresponding to the only active variables ϑ_N, φ_N, and ϕ_N, respectively. All other quantum numbers have zero values, as seen from Eq. (7.4); corresponding variables do not appear in the PH. Thus, if we use the PH basis to expand a general function in the $3N$-dimensional space, variables associated with zero quantum numbers remain 'frozen.' Clearly, this subset is useful in a situation in which physical condition asserts that there is no or at most a weak dependence on the variables associated with zero quantum numbers. If a quantum system has no correlations higher than two-body correlations, then, the PH basis is proper for the expansion of the wave function of the system. The Bose–Einstein condensate achieved in the laboratory is a good example of such a system. We will discuss its use in Chap. 8. The degeneracy of a general HH of order L increases very rapidly with L, corresponding to all allowed sets of quantum numbers $\{(l_1, m_1), (n_2, l_2, m_2), (n_3, l_3, m_3), \ldots, (n_N, l_N, m_N)\}$. If the total orbital angular momentum (l) is good for the system, then the degeneracy of a PH of order $2K + l$ is only $2l + 1$. Thus the effort is reduced greatly, when the PH basis is used.

The orthonormalization satisfied by the PH is obtained from that of the general HH, and is given by

$$\int \mathcal{P}_{2K+l}^{l,m}(\Omega_N^{(ij)}) \, \mathcal{P}_{2K'+l'}^{l',m'}(\Omega_N^{(ij)}) \, d\Omega_N^{(ij)} = \delta_{KK'} \, \delta_{ll'} \, \delta_{mm'}. \tag{7.7}$$

7.2 Potential Multipoles

We next discuss how a general two-body potential $V(\vec{r}_{ij})$ can be expanded in the corresponding set of PH $\{\mathcal{P}_{2K+\lambda}^{\lambda,\mu}(\Omega_N^{(ij)})\}$. The potential (specially in nuclear systems) generally contains a central plus a tensor term with dependence on spin and isospin operators. Hence, we can write

$$V(\vec{r}_{ij}) = \sum_{\lambda,\mu} A_{\lambda\mu}(i,j) Y_{\lambda\mu}(\hat{r}_{ij}) V_\lambda(r_{ij}), \tag{7.8}$$

where $A_{\lambda\mu}(i,j)$ is an operator acting on spin and isospin variables of the pair (ij), but independent of r_{ij}. λ is the rank of the spherical tensor represented by the potential and μ (which can take values $-\lambda, -\lambda+1, \ldots, \lambda-1, \lambda$) is its component, e.g., λ is zero for a central potential and has the value 2 for a rank 2 tensor potential, as is common in the case of nuclear interactions. The quantity $V_\lambda(r_{ij})$ is the corresponding radial dependence of the potential. The PH basis is sufficient for an expansion of $V(\vec{r}_{ij})$ in the full HH basis. Expanding $V_\lambda(r_{ij}) Y_{\lambda\mu}(\hat{r}_{ij})$ in $\{\mathcal{P}_{2K''+\lambda}^{\lambda,\mu}(\Omega_N^{(ij)})\}$, we finally have

$$V(\vec{r}_{ij}) = \sum_{K'',\lambda,\mu} \mathcal{P}_{2K''+\lambda}^{\lambda,\mu}(\Omega_N^{(ij)}) V_{K''}^{(3N,\lambda)}(\xi) A_{\lambda,\mu}(i,j), \tag{7.9}$$

where $V_{K''}^{(3N,\lambda)}(\xi)$ is called the potential multipole. Using Eq. (7.5) it is given by

$$
\begin{aligned}
V_{K''}^{(3N,\lambda)}(\xi) &= \langle \mathcal{P}_{2K''+\lambda}^{\lambda,\mu}(\Omega_N^{(ij)}) | V(\vec{r}_{ij}) \rangle \\
&= \mathcal{Y}_{[0]}(3N-3) \int_0^{\pi/2} {}^{(N)}\mathcal{P}_{2K''+\lambda}^{\lambda,0}(\phi) V_\lambda(r\cos\phi)(\sin\phi)^{3N-4}\cos^2\phi \, d\phi.
\end{aligned}
\tag{7.10}
$$

We have displayed the spin–isospin operator $A_{\lambda\mu}(i,j)$ of $V(\vec{r}_{ij})$ explicitly in Eq. (7.9). Note that the potential multipole does not depend on the interacting-pair label (ij), since the former is a function of ξ only, which is invariant under all permutations of particle labels. This dependence is taken over by the potential harmonic in Eq. (7.9). Expanding the two-body potential as in Eq. (7.9), one can calculate the potential matrix elements in terms of geometrical structure coefficients $\langle \mathcal{P}^{l,m}_{2K+l}(\Omega^{(ij)}_N)|$ $\mathcal{P}^{\lambda,\mu}_{2K''+\lambda}(\Omega^{(kl)}_N)|\mathcal{P}^{l',m'}_{2K'+l'}(\Omega^{(i'j')}_N)\rangle$ (where the $\langle \cdots \rangle$ indicates integration over the hyperangles in the $3N$-dimensional space) and potential multipoles $V^{(3N,\lambda)}_{K''}(\xi)$. Dependence on the pair labels (ij), $(i'j')$ and (kl) is taken up in the next section.

7.3 Overlap of PHs of Different Pairs

The PH depends on the interacting-pair label (ij). In the CDE and the potential matrix element such labels will enter. In this section, we discuss how it can be handled. The general overlap of two PH elements has the form $\langle \mathcal{P}^{l,m}_{2K+l}(\Omega^{(ij)}_N)|\mathcal{P}^{l',m'}_{2K'+l'}(\Omega^{(i'j')}_N)\rangle$. Since the grand orbital $2K+l$, as also orbital angular momentum quantum numbers l, m are conserved under renaming of particle labels, this overlap vanishes unless $K' = K$, $l' = l$, and $m' = m$. Now one of the particle pairs can be chosen arbitrarily as (12). Then, the only nonvanishing overlaps of interest are $\langle \mathcal{P}^{l,m}_{2K+l}(\Omega^{(12)}_N)|\mathcal{P}^{l,m}_{2K+l}(\Omega^{(ij)}_N)\rangle$. Three situations can arise: when both pair of labels are the same [i.e., when $(ij) = (12)$], when one of the labels is the same [i.e., when only one of (ij) is greater than 2], and when there is no common label [i.e., when both $i, j > 2$]. It can be seen [1] that

$$\langle \mathcal{P}^{l,m}_{2K+l}(\Omega^{(12)}_N)|\mathcal{P}^{l,m}_{2K+l}(\Omega^{(ij)}_N)\rangle = \frac{{}^{(N)}\mathcal{P}^{l,0}_{2K+l}(\varphi^{(ij)}_N)}{{}^{(N)}\mathcal{P}^{l,0}_{2K+l}(0)}$$

$$= (\cos \varphi^{(ij)}_N)^l \frac{P^{(3N-5)/2,l+\frac{1}{2}}_K(\cos 2\varphi^{(ij)}_N)}{P^{(3N-5)/2,l+\frac{1}{2}}_K(1)}, \quad (7.11)$$

where

$$\varphi^{(ij)}_N = 0 \quad \text{for } (ij) = (12)$$
$$= \tfrac{\pi}{2} \quad \text{for } i, j > 2$$

and $\cos \varphi^{(ij)}_N = \pm \frac{1}{2}$, for equal mass particles with one common label. The simplest case is when $(ij) = (12)$. In this case we see that the overlap is 1, as expected. For $i, j > 2$, i.e., no common index, we get from Eq. (7.11)

$$\langle \mathcal{P}^{l,m}_{2K+l}(\Omega^{(12)}_N)|\mathcal{P}^{l,m}_{2K+l}(\Omega^{(ij)}_N)\rangle = \frac{(-1)^K}{2^{2K}} \frac{(2K+1)!}{K!} \frac{\Gamma((3N-3)/2)}{\Gamma(K+(3N-3)/2)} \delta_{l0}.$$
$$(i, j > 2) \quad (7.12)$$

When there is only one common index (one of i, j is greater than 2, the other is less than or equal to 2), one has to use Eq. (7.11) with $\cos \varphi^{(ij)}_N = \pm \frac{1}{2}$.

7.4   Potential Basis as Optimal Subset

Calculation of the potential matrix element of different interacting pairs is a formidable task, if the entire HH basis is used. The labor in handling this can be reduced by the use of optimal subset. This is particularly convenient if the first term (which is of order zero and independent of all hyperangles) of HH expansion of two-body potential is the dominant term. Then, the HH expansion of the many-body wave function will also have a dominant contribution from the HH of order zero, $viz.$, $\mathcal{Y}_{[0]}(3N)$. Hence only the PH subset will be the subset of HH, which is directly connected with the dominant HH (of order zero) of the wave function, through the two-body potential. Then according to Eq. (4.36), the optimal subset will be the PH basis. It will also be referred to as the *potential basis*. Since according to this assumption, the dominant part of the potential is independent of the hyperangles, coupling of various HH components of the wave function mediated through the two-body potential can be treated as a small perturbation. The validity of the assumption that the zeroth order HH contributes dominantly has been tested for few nucleon system interacting via various standard nucleon–nucleon potentials. Contribution of the $K = 0$ term to the ground state energy has been found to be 80 % or more [1].

7.4.1   Symmetrical PH Basis

The PH basis can be symmetrized according to the identity and type of the particles. Simplest to construct is the totally symmetric space state. For the ground state of a system of A identical bosons, all the particles are likely to be in the lowest single particle state. Thus, the lowest order HH with $L = 0$ (i.e., $l = m = 0$, $K = 0$) is dominant. Even for a system of fermions (like the nucleus), the space totally symmetric state has a dominant contribution. An example is provided by the trinucleon ground state (see Chap. 5). An element of the optimal subset comprised of symmetrical combination of PH is

$$B_{2K}^{(S)}(\Omega_N) = C_K \sum_{i,j>i}^{A} \mathcal{P}_{2K}^{0,0}(\Omega_N^{(ij)}), \tag{7.13}$$

where C_K is a normalization constant and $\mathcal{P}_{2K}^{0,0}(\Omega_N^{(ij)})$ is the PH, given by Eq. (7.5), corresponding to $l = m = 0$. The elements of this basis is totally symmetric under any pair exchange, and hence it is called the *symmetrical potential basis*. The ground state wave function for the relative motion can be expanded as

$$\Psi_{gs}(\xi, \Omega_N) = \xi^{-(3N-1)/2} \sum_{K=0}^{\infty} u_K(\xi) B_{2K}^{(S)}(\Omega_N). \tag{7.14}$$

The system of CDE resulting from this expansion is

$$
\left[-\frac{\hbar^2}{m}\frac{d^2}{d\xi^2} + \frac{\hbar^2}{m}\frac{\mathcal{L}_K(\mathcal{L}_K+1)}{\xi^2} - E \right] u_K(\xi)
$$
$$
+ \frac{A(A-1)}{2}\sum_{K'}\langle B^{(S)}_{2K}|V(\vec{r}_{ij})|B^{(S)}_{2K'}\rangle u_{K'}(\xi) = 0, \tag{7.15}
$$

where $\mathcal{L}_K = 2K + \frac{3N-3}{2}$. Since the expansion basis is totally symmetric under any pair exchange, the matrix element of sum of all interacting pairs is just the matrix element of one pair multiplied by the number of pairs. Thus handling of all pair-wise interaction becomes much simpler. However, the calculation of the symmetrical potential basis is not so easy. Detailed calculation of the potential matrix element in this case, can be found in Ref. [1]. In the next section, we discuss how the potential matrix element can be calculated in the unsymmetrical PH basis.

7.5 Potential Matrix in Unsymmetrized PH Basis

We can also use the unsymmetrized PH basis, given by Eq. (7.5), say for the particular choice of $(ij) = (12)$, i.e., the set $\{\mathcal{P}^{l,m}_{2K+l}(\Omega^{(12)}_N)\}$. In this case, we cannot replace the matrix element of sum of all pair-wise potentials as the number of pairs times the matrix element of any pair potential. Expanding the ground state wave function in the unsymmetrized basis with $l = m = 0$

$$
\Psi_{gs}(\xi, \Omega_N) = \xi^{-(3N-1)/2}\sum_{K'=0}^{\infty} u_{K'}(\xi)\mathcal{P}^{0,0}_{2K'}(\Omega^{(12)}_N). \tag{7.16}
$$

Substitution of this in the Schrödinger equation and projection on $\mathcal{P}^{0,0}_{2K}(\Omega^{(12)}_N)$ results in the set of CDE

$$
\left[-\frac{\hbar^2}{m}\frac{d^2}{d\xi^2} + \frac{\hbar^2}{m}\frac{\mathcal{L}_K(\mathcal{L}_K+1)}{\xi^2} - E \right] u_K(\xi)
$$
$$
+ \sum_{K'}\left[\langle\mathcal{P}^{0,0}_{2K}|\sum_{i,j<i}^{A}V(\vec{r}_{ij})|\mathcal{P}^{0,0}_{2K'}\rangle\right] u_{K'}(\xi) = 0. \tag{7.17}
$$

Expanding the potential as in Eq. (7.9), we have, in general

$$
V_{KK'}(\xi) \equiv \langle\mathcal{P}^{l,m}_{2K+l}|\sum_{i,j<i}^{A}V(\vec{r}_{ij})|\mathcal{P}^{l',m'}_{2K'+l'}\rangle
$$

$$= \sum_{K''\lambda\mu} V_{K''}^{(3N,\lambda)}(\xi) \sum_{i,j<i}^{A} \langle A_{\lambda\mu}(ij) \rangle_{\text{spin-isospin}}$$

$$\times \langle \mathcal{P}_{2K+l}^{l,m}(\Omega_N^{(12)}) | \mathcal{P}_{2K''+\lambda}^{\lambda,\mu}(\Omega_N^{(ij)}) | \mathcal{P}_{2K'+l'}^{l',m'}(\Omega_N^{(12)}) \rangle. \qquad (7.18)$$

The last factor is the geometrical structure coefficient, introduced in Chap. 3, although in the present case, the PH in the middle corresponds to a different partition. Since the global quantum numbers L, l and m are conserved under all permutations of particles, $\mathcal{P}_{2K''+\lambda}^{\lambda,\mu}(\Omega_N^{(ij)})$ can be expanded in a corresponding set of PH belonging to partition (12). We will specialize to the ground state of the system with $l = m = 0$, $l' = m' = 0$ and scaler interaction ($\lambda = \mu = 0$) only. Then the PH $\mathcal{P}_{2K''}^{0,0}(\Omega_N^{(ij)})$ can be expressed as a linear combination of $\mathcal{P}_{2\chi}^{0,0}(\Omega_N^{(12)})$ as

$$\mathcal{P}_{2K''}^{0,0}(\Omega_N^{(ij)}) = \sum_{\chi} c_{K''\chi} \mathcal{P}_{2\chi}^{0,0}(\Omega_N^{(12)}) \qquad (7.19)$$

Since the global quantum number L is conserved, we must have $\chi = K''$. Hence

$$\mathcal{P}_{2K''}^{0,0}(\Omega_N^{(ij)}) = c_{K''K''} \mathcal{P}_{2K''}^{0,0}(\Omega_N^{(12)})$$

$$= \langle \mathcal{P}_{2K''}^{0,0}(\Omega_N^{(12)}) | \mathcal{P}_{2K''}^{0,0}(\Omega_N^{(ij)}) \rangle \mathcal{P}_{2K''}^{0,0}(\Omega_N^{(12)}). \qquad (7.20)$$

The first factor can be obtained from Eq. (7.11). For simplicity, we consider a scalar interaction, independent of spin–isospin operators, for which $A_{\lambda,\mu}(i,j)$ becomes an identity operator. The potential matrix element appearing in Eq. (7.18) becomes

$$V_{KK'}(\xi) = \sum_{K''} \left[\left(\sum_{i<j}^{A} \langle \mathcal{P}_{2K''}^{0,0}(\Omega_N^{(12)}) | \mathcal{P}_{2K''}^{0,0}(\Omega_N^{(ij)}) \rangle \right) \right.$$

$$\left. \times V_{K''}^{(3N,0)}(\xi) \langle \mathcal{P}_{2K}^{0,0}(\Omega_N^{(12)}) | \mathcal{P}_{2K''}^{0,0}(\Omega_N^{(12)}) | \mathcal{P}_{2K'}^{0,0}(\Omega_N^{(12)}) \rangle \right]. \qquad (7.21)$$

The GSC appearing above is the standard one, defined as in Chap. 3, for the principal partition (12) chosen. Its evaluation can also be done by the LIE technique explained there. The dependence on the labels of all interacting pairs is given through

$$f_{K'',l=0}^{2} \equiv \sum_{i<j}^{A} \langle \mathcal{P}_{2K''}^{0,0}(\Omega_N^{(12)}) | \mathcal{P}_{2K''}^{0,0}(\Omega_N^{(ij)}) \rangle.$$

Calculation of f_{Kl}^2

Consider, the general case

$$f_{Kl}^2 = \sum_{i<j}^{A} \langle \mathcal{P}_{2K}^{l,0}(\Omega_N^{(12)}) | \mathcal{P}_{2K}^{l,0}(\Omega_N^{(ij)}) \rangle. \tag{7.22}$$

To evaluate the sum over different partitions in Eq. (7.22), we note that partition (ij) has three possibilities:

1. Both labels are the same [only one possibility $(i, j) = (1, 2)$]
   Clearly

$$\langle \mathcal{P}_{2K}^{l,0}(\Omega_N^{(12)}) | \mathcal{P}_{2K}^{l,0}(\Omega_N^{(12)}) \rangle = 1.$$

   This also follows from Eq. (7.11) with $\varphi_N^{(12)} = 0$.

2. (i, j) with one common label with $(1, 2)$
   There are two possibilities: $i = 1, j > 2$ and $i = 2, j > 2$. Both have $(A - 2)$ contributions to the sum with $j = 3, \ldots, A$. In this case $\varphi_N^{(31)} = \pm\frac{1}{2}$ and we have

   contribution from one common label

$$= 2(A - 2)\langle \mathcal{P}_{2K}^{l,0}(\Omega_N^{(12)}) | \mathcal{P}_{2K}^{l,0}(\Omega_N^{(31)}) \rangle = 2(A - 2)\left(\frac{1}{2}\right)^0 \frac{P_K^{\alpha,\beta}(-\frac{1}{2})}{P_K^{\alpha\beta}(1)}$$

$$= 2(A - 2)\frac{P_K^{\alpha,\beta}(-\frac{1}{2})}{P_K^{\alpha\beta}(1)},$$

   where $\alpha = (3N - 5)/2$ and $\beta = l + \frac{1}{2}$.

3. No common labels, $i > 2, j > i$. We can take $(i, j) = (3, 4)$. There are $^{(A-2)}C_2$ such contributions with $\varphi_N^{(34)} = \frac{\pi}{2}$, giving

   contribution from no common indices

$$= \frac{(A - 2)(A - 3)}{2} \langle \mathcal{P}_{2K}^{l,0}(\Omega_N^{(12)}) | \mathcal{P}_{2K}^{l,0}(\Omega_N^{(34)}) \rangle$$

$$= \frac{(A - 2)(A - 3)}{2} (0)^l \frac{P_K^{\alpha,\beta}(-1)}{P_K^{\alpha\beta}(1)}$$

$$= \delta_{l0} \frac{(A - 2)(A - 3)}{2} \frac{P_K^{\alpha,\beta}(-1)}{P_K^{\alpha\beta}(1)}$$

Combining the contributions of the three possibilities in Eq. (7.22), we finally have [2]

$$f_{Kl}^2 = 1 + [2(A - 2)P_K^{\alpha\beta}(-\frac{1}{2}) + \delta_{l0}\frac{(A - 2)(A - 3)}{2}P_K^{\alpha\beta}(-1)]/P_K^{\alpha\beta}(1). \tag{7.23}$$

Use of the overlap f_{Kl}^2 reduces the algebraic complexity of considering all pairs separately. Its use in the Faddeev equation will be presented in Chap. 8. A similar approach has been used in computing the projection function for the S states in the integro-differential equation approach [3].

Dominance of two-body correlation in the many-body wave function is the basis of the potential harmonic approximation. Contribution of such two-body correlations to the wave function of a many-body system was investigated by Fabre et al [4]. They studied self conjugate closed shell nuclei ^{4}He, ^{16}O and ^{40}Ca with the PH basis and compared the calculated binding energies with those of the full set. The results agree closely, showing the importance of two-body correlations. The PH basis is appropriate for the dilute Bose–Einstein condensates, which will be taken up in Chap. 8. Yalcin and Smisek used the PH approximation in atomic three-body systems [5].

References

1. Fabre de la Ripelle, M.: Ann. Phys. (N.Y.) **147**, 281 (1983)
2. Fabre de la Ripelle, M.: Few-Body Syst. **1**, 181 (1986)
3. Fabre de la Ripelle, M., Fiedeldey, H., Sofianos, S.A.: Phys. Rev. C **38**, 449 (1988)
4. Fabre de la Ripelle, M., Ballot, J.L., Navarro, J.: Phys. Lett. B **143**, 19 (1984)
5. Yalcin, Z., Simsek, M.: Int. J. Quan. Chem. **88**, 735 (2002)

Chapter 8
Application to Bose–Einstein Condensates

Abstract Bose–Einstein condensate (BEC) is formed when a macroscopic fraction of bosons in a Bose gas occupies the lowest energy state, below a critical temperature. It is extremely dilute and the effective two-body interaction is given in terms of s-wave scattering length (a_s). Properties of BEC are discussed. The standard Gross–Pitaevskii equation (GPE) is obtained from mean field theory with contact interaction. Simplifying assumptions and limitations of GPE are discussed. Many-body treatment consists of solving the many-body equation by expanding the interacting pair Faddeev component in correlated PH basis (PHEM), which is appropriate for the dilute system. For the extremely dilute system interacting via van der Waals potential, a short-range correlation function is needed with the PH basis. Results for attractive and repulsive condensates are presented.

In 1924 S.N. Bose explained the black body radiation introducing the Bose distribution function for photons, which are massless, spin one particles [1]. Einstein extended the idea to massive bosons (integral spin particles) and predicted that a macroscopic fraction of particles in such a bosonic noninteracting gas can occupy the lowest energy state below a critical temperature [2]. This state of matter is called Bose–Einstein condensate (BEC). Details of the BEC can be found in Refs. [3–5]. It is obvious that this can occur at a very low temperature (estimated to be of the order of 10–100 nano-Kelvin), so that thermal excitations cannot scatter bosons into higher energy levels. At the critical temperature the thermal de Broglie wavelength becomes comparable to the average interparticle separation. Another condition to achieve BEC is that the effect of two-body interactions must be quite small, since such interactions can also populate higher energy levels. Thus the bosons must form an *extremely dilute* cloud, so that average interparticle separation is much larger than the range of interparticle interaction. This condition gives the number density of a typical laboratory BEC to be $\sim 10^{13}$–$10^{15}\,\mathrm{cm}^{-3}$. This is extremely small compared to the density of molecules in air at room temperature ($\approx 10^{19}\,\mathrm{cm}^{-3}$). The diluteness prevents the cloud from condensing to the liquid or solid state, since three-body collisions (which are necessary to form bound molecules or clusters, in order that the third particle can escape with the released binding energy) would be practically absent.

© Springer India 2016
T.K. Das, *Hyperspherical Harmonics Expansion Techniques*,
Theoretical and Mathematical Physics, DOI 10.1007/978-81-322-2361-0_8

Even though BEC was predicted in 1924, its experimental achievement in dilute gases was not possible until 1995, since such a low temperature could not be reached by conventional cryogenic means. Moreover one has to find a container, which can hold the condensate, without heating it up. Both cooling and confinement was successfully done using very clever laser techniques and magneto-optic confinement. The experimental observation of BEC has renewed a great deal of interest in it— experimental as well as theoretical. The importance of this topic is clearly demonstrated by the fact that three independent Nobel Prizes were awarded on BEC related works in quick succession in the recent past.

In 1938, London suggested that the superfluidity of liquid Helium is related to BEC. But the average interparticle separation in the liquid phase is of the order of interatomic forces. Hence atoms in the lowest energy state are scattered into higher levels and the occupation number in the lowest energy state is reduced even at zero temperature.

Attempts were made during the 1980s and 1990s to create BEC in extremely dilute Bose gases, so that the average interparticle separation is much larger than the range of interparticle interactions. This reduces the probability of molecule formation to practically zero. As mentioned above, molecule formation is possible only through three and higher body collisions, so that two atoms form a molecule, while the third takes away the released binding energy as its kinetic energy. To get an idea of the temperature (T) needed to achieve BEC, we can compare the thermal de Broglie wavelength, $\lambda_T \sim \sqrt{\hbar^2/(2mk_B T)}$ (where k_B and m are the Boltzmann constant and mass of the atom, respectively) with the mean interatomic separation, $n^{-\frac{1}{3}}$ (n being the number density). Since n is in the range of 10^{13}–10^{15} cm^{-3}, T turns out to be in the range from 100 nK to a few μK. Two main experimental difficulties were: (1) to attain extremely low temperatures (typically a few hundred nano-Kelvin) and (2) to find a container for the gas. Any material container will involve collisions with the walls of the container. This will lead to loss (due to absorption by the wall) and heating. Very low temperatures are obtained by laser cooling followed by evaporative cooling [3, 4]. The container problem is relatively easily solved by confining the gas in magnetic and optical traps. The trap potential is usually chosen to be that of a harmonic oscillator of frequency ω (typically ~100 Hz). This gives the BEC length scale $a_{\mathrm{ho}} = \sqrt{\frac{\hbar}{m\omega}}$.

Bose–Einstein condensation was first achieved experimentally in gases of neutral alkali-atom (^{87}Rb, ^{23}Na and ^{7}Li) vapors in 1995 [3, 4]. The total number of fermions (protons, neutrons, and electrons) in such an atom is even. Hence they behave as bosons. In BEC experiments, the apparatus is at room temperature, while the atoms are trapped and cooled to incredibly low temperatures. Atoms are initially laser cooled to about 10 μK by applying laser beams in six directions, counter propagating pairs along the three Cartesian axes. The laser frequency is chosen in such a way that the moving atom will be in resonance *only* when it encounters a photon coming from the opposite direction and *not* in the same direction. This is done by slightly detuning the laser frequency, such that the Doppler-shifted frequency, as seen by an approaching atom, is in resonance. Hence only the photon coming from the opposite direction is selectively absorbed. This slows down the atom and thus reduces the temperature of the atomic cloud. The excited atom later spontaneously emits a photon. Since the process is random and the photon can be emitted in any direction, there is no net gain in momentum in the spontaneous emission process. Evaporative cooling is next used, in which external magnetic fields are manipulated to expel from the trap the most energetic among the remaining atoms, thereby reducing the temperature to the desired value. Detection of BEC is done as follows. After reaching a desired temperature, the confining trap is switched off, allowing the atoms to move outward freely. Fastest atoms move furthest. After a time lag, the expanding cloud is imaged by a probe laser beam. The density profile corresponds to a velocity distribution at the time when the trap was switched off. At higher temperatures, a Gaussian distribution (corresponding to the thermal distribution) is observed. A sharp peak above the Gaussian pedestal in the velocity distribution was observed below a certain critical temperature (T_c). Temperature of the cloud is determined by fitting the Gaussian pedestal with the thermal distribution at a temperature T. The observed sharp peak at the center indicates an appreciable number of atoms with practically zero speed and hence provides a clear signature for the experimental attainment of BEC.

The size of the cloud is $\sim a_{ho}$ and it is a macroscopic length. It can be a few tenths of a millimeter. Hence the structure of the condensate wave function can be investigated directly by optical means, as discussed above. Below T_c, most of the particles are in the ground state of the harmonic trap. Hence single particle wave functions of these atoms overlap and the many-body system behaves as a single quantum entity, where the atoms move in a coherent fashion, describable in terms of a single macroscopically extended wave function, called the 'condensate wave function.' This provides a unique opportunity to explore quantum phenomena on a macroscopic scale. Interatomic interactions can produce a measurable effect on the condensate and are of great interest. The wave function is a product of single particle wave functions and has no correlations (for definition of correlations see Chap. 4, Sect. 4.5 and Chap. 7). On the other hand, inclusion of two-body interactions makes it a nontrivial many-body problem. In the following sections, we will discuss different theoretical approaches to the problem.

8.1 General Properties of BEC

From the above discussion, we enlist some basic features of a typical laboratory BEC:

1. The harmonic oscillator trap in a laboratory BEC has frequency typically ~ 100 Hz, corresponding to a length scale, $a_{ho} \approx 2 \times 10^4 \, a_0$ for the Rb condensate, where a_0 is the Bohr radius (atomic unit of length), $a_0 = 0.529 \times 10^{-8}$ cm. Thus the BEC length scale is four orders of magnitude larger than the atomic length scale.
2. Average interparticle separation in the BEC, $r_{av} = n^{-\frac{1}{3}} \sim 2 \times 10^{-5}$ cm $\approx 4 \times 10^3 \, a_0$. Thus the average separation is also much larger than the atomic length scale.

3. BEC was originally considered for noninteracting bosons, for which the many-body wave function is a product of single particle wave functions. But actual bosonic atoms interact among themselves. Typical energy of a bosonic atom in the BEC is $\sim \hbar \omega$. For $\omega = 100\,\mathrm{Hz}$, this is $\sim 10^{-13}\,\mathrm{eV}$. It is negligibly small compared to atomic energy scale (eV). This shows that the atoms in the BEC scatter with practically zero energy. Hence the scattering cross section is $4\pi |a|^2$, where a is the s-wave scattering length. At such low energies, the interacting pair of atoms are too far apart to feel the bare (actual) interatomic interaction (which is a van der Waals type attraction, $-\frac{C_6}{r_{ij}^6}$, r_{ij} being the interacting pair separation), having a range of $\sim 20\,a_0$. Thus, the effective interaction between the pair is controlled entirely by the s-wave scattering. Value of a is about $100\,a_0$ for a condensate of ^{87}Rb atoms. In the zero energy limit, the effective interaction in the momentum space [4] becomes a constant $U_0 = \frac{4\pi \hbar^2 a}{m}$. Hence in coordinate space, it becomes a contact (delta function) interaction of strength U_0

$$V_{\text{effective}}(r_{ij}) = \frac{4\pi \hbar^2 a}{m} \delta(r_{ij}). \tag{8.1}$$

This *effective interaction* can be attractive or repulsive accordingly as a is < 0 or > 0 respectively, whereas the bare interatomic interaction is *always attractive* at separations greater than the repulsive core ($\sim 5\,a_0$) of the actual interatomic interaction. Also, since the scattering cross section due to a hard sphere of radius a is $4\pi a^2$, we may take the sphere of interaction around an atom to be a sphere of radius $|a|$. For a very dilute system for which r_{av} is much larger than the range of bare interatomic interaction and at negligible energy, the effective interaction (and *not* the bare interaction) will give correct results in the first Born approximation for scattering and mean field theory for bound states [4].

4. Since r_{av} is quite large compared to the radius of influence of the effective interaction $|a|$, the condensate wave function can be taken as having no correlations, in the lowest approximation. Hence the wave function becomes a product of single particle wave functions and the mean field approximation is a reasonable one. This approximation together with Eq. (8.1) gives rise to the mean-field Gross–Pitaevskii (GP) equation (Sect. 8.2). However, since two-body scattering generates the effective two-body interaction, pair correlations play an important role. This is also demonstrated by the fact that although $|a|$ is small, it is not very small compared to r_{av}. For larger $|a|$, correlations higher than two-body ones will become relevant.

5. One of the conditions for achieving BEC in the laboratory is that three and higher body collisions must be absent, in order that there is no depletion of the condensate. This means that the number of particles within the sphere of interaction must be small, i.e., $n|a|^3 \ll 1$. This implies that three-body and higher body correlations are absent. We will see in the following that it will simplify the many-body treatment immensely (Sect. 8.3).

8.2 GP Equation

Substitution of Eq. (8.1) in the mean field equation gives

$$\left[-\frac{\hbar^2}{2m}\nabla^2 + \frac{1}{2}m\omega^2 r^2 + U_0|\psi(\vec{r})|^2\right]\psi(\vec{r}) = \mu\psi(\vec{r}),\qquad(8.2)$$

where $U_0 = \frac{4\pi\hbar^2 a}{m}$ and μ is the chemical potential. $\psi(\vec{r})$ is the condensate wave function, normalized as

$$\int|\psi(\vec{r})|^2 d\vec{r} = N,\qquad(8.3)$$

where N is the number of bosonic atoms in the condensate (in this section, we adopt the symbols N and a for number of particles and the s-wave scattering length, respectively, to be consistent with the literature). The density of particles is

$$n(\vec{r}) = |\psi(\vec{r})|^2.\qquad(8.4)$$

$\psi(\vec{r})$ can be written as

$$\psi(\vec{r}) = N^{\frac{1}{2}}\phi(\vec{r}),\qquad(8.5)$$

where $\phi(\vec{r})$ is the single particle wave function, normalized as

$$\int|\phi(\vec{r})|^2 d\vec{r} = 1.\qquad(8.6)$$

Equation (8.2) is the celebrated GP equation, obtained independently by Gross and Pitaevskii in 1961. The above outline is an intuitive and heuristic argument leading to the GP equation, but not a rigorous derivation. Moreover, a rigorous definition of contact interactions in more than one dimensions is not straightforward (see, e.g., [6, 7] and references quoted therein), and both mathematically and physically it is more appropriate to start with a bona fide interaction potential. Limitations of the intuitive arguments are discussed in the following subsection. On the other hand, since the laboratory achievement of BEC in 1995, a wealth of accurate experimental results have become available. Most of them agree well with the predictions of the GP equation. Thus the GP equation should have a stronger basis than that presented above. Indeed mathematically rigorous works have been carried out on BEC and the GP equation over the last 15 years, leading to a mathematical proof of existence of BEC in the GP limit [8]. We will briefly discuss them in Sect. 8.2.2.

The GP equation is a nonlinear Schrödinger equation, the nonlinearity comes from the interaction term proportional to particle density $n(\vec{r}) = |\psi(\vec{r})|^2$. As we discussed earlier, this equation ignores all correlations (at least on length scales beyond the scattering length) and is expected to be valid in the *dilute* limit $n|a|^3 \ll 1$. It has been very successful in explaining many observed quantities in dilute BEC [3]. Except for its nonlinearity, the GP equation is a simple one. It can be studied by variational methods [3]. Accurate numerical codes have been developed for the numerical solution, see for example Ref. [9]. We will not discuss applications of the GP equation, as our main aim is to introduce the PH expansion to the many-body treatment of BEC. Interested readers can find details of the GP equation and its applications in Refs. [3, 4].

8.2.1 Simplifying Assumptions and Their Limitations

In spite of the overwhelming success of the GP equation, the heuristic arguments of Sect. 8.2 have limitations. Let us list the simplifying assumptions and limitations of their validity:

1. The fundamental assumption is that the condensate is very dilute, $n|a|^3 \ll 1$. For a condensate of 10^6 atoms of ^{87}Rb ($a = 100\, a_0$) in a harmonic trap of frequency $\omega = 100\,$Hz, the diluteness parameter $n|a|^3 = 0.081$ and the diluteness condition is fairly well satisfied. For smaller number of atoms, it is well satisfied, but for 10^8 atoms, the diluteness parameter is larger than 1 and the condition is not satisfied.
2. Recent experimental setups using Feshbach resonance can change a for certain atoms (e.g., ^{85}Rb) to almost any value. Thus even for a tenfold increase $a = 1000a_0$, GP equation is not valid even for 10^5 atoms. In such cases, correlations enter into the wave function (see Sect. 8.3).
3. Assumption of a contact interaction, Eq. (8.1) is not physically *bona fide*, but it is a reasonable approximation when the diluteness condition is satisfied. However, this is not a good assumption for excited states of the condensate with higher energies, as also for observables which are sensitive to the details of the two-body interactions. In such cases, an effective interaction derived from a bare realistic interatomic interaction is desirable (see Sect. 8.3).
4. Assumption of a mean field approach is only an approximation and it is again justified for the extremely dilute condensate only. A better approach would be a many-body treatment using realistic two-body interactions and including correlations.

In the next subsection, we will briefly outline the rigorous mathematical treatment of BEC and the GP equation. In Sect. 8.3, we will see how some of the limitations listed above can be addressed by the application of PH expansion method in an approximate many-body treatment, even if the condensate is not very dilute.

8.2.2 Rigorous Proof of Existence of BEC and Derivation of the GP Equation

BEC was originally proposed for noninteracting bosons, for which the many-body ground state is a product of single particle wave functions and all the bosons occupy this state. But realistic bosonic atoms have interactions and the ground state wave function is not strictly a product wave function. Thus a rigorous proof of existence of BEC in interacting bosons is needed.

Furthermore, the heuristic argument presented at the beginning of Sect. 8.2 to arrive at the GP equation is not a rigorous derivation. This is because the mean field theory assumes the many-body wave function to be a product of single particle wave functions, which is rigorously valid only for noninteracting particles. In particular, if the two-body interaction has a hard core, the expectation value of the Hamiltonian diverges [8]. For a finite result, the product wave function must be multiplied by a Jastrow type function $F(\vec{r}_1, \ldots, \vec{r}_A)$ [10], which vanishes whenever any pair separation becomes less than or equal to the hard core radius. The function F should be chosen judiciously involving the zero-energy scattering solution of the two-body problem [11]. Moreover, the contact interaction (8.1) is not a *bona fide* potential. It needs to be replaced by a physical potential having a finite range.

A rigorous mathematical proof of BEC for dilute trapped Bose gases of atoms interacting through repulsive two-body interactions and derivation of the Gross–Pitaevskii energy functional were provided by Lieb, Seiringer, Yngvason, and other authors. We will briefly outline the steps in the following. Interested readers should read the original papers cited below or detailed lecture notes [8, 11–14].

The first rigorous proof of BEC in a gas-containing N bosonic atoms trapped by an external potential $V(\vec{r}_i)$ and interacting via two-body repulsive interaction $v(\vec{r}_i - \vec{r}_j)$ in the dilute limit was given by Lieb and Seiringer [15], starting from Schrödinger equation for the many-body Hamiltonian

$$H = \sum_{i=1}^{N} \left[-\nabla_i^2 + V(\vec{r}_i) \right] + \sum_{1 \leq i < j \leq N} v(\vec{r}_i - \vec{r}_j). \tag{8.7}$$

In this subsection, the units are so chosen that $\frac{\hbar^2}{2m} = 1$. The interaction potential v is assumed to be spherically symmetric, nonnegative, and having scattering length a, which is allowed to vary with N. This is done by scaling: writing $v(\vec{r}) = v_1(\vec{r}/a)/a^2$, where v_1 is kept fixed when varying a and it has scattering length 1. The ground state properties of such a trapped dilute, repulsively interacting gas are usually described in the GP theory by means of the Gross–Pitaevskii energy functional, given by

$$\mathcal{E}^{\mathrm{GP}}[\phi] = \int \left[|\vec{\nabla}\phi(\vec{r})|^2 + V(\vec{r})|\phi(\vec{r})|^2 + g|\phi(\vec{r})|^4 \right] d^3r, \tag{8.8}$$

where $g = 4\pi N a$. Let ϕ^{GP} minimize $\mathcal{E}^{\mathrm{GP}}$, subject to normalization condition $\int |\phi^{\mathrm{GP}}|^2 d^3r = 1$, corresponding energy being $E^{\mathrm{GP}}(g)$ (which is the lowest value of $\mathcal{E}^{\mathrm{GP}}$). The asymptotic exactness of the GP approximation for the ground state of such a gas was established rigorously by Lieb, Seiringer, and Yngvason [15, 16], where it was shown that the GP energy functional correctly describes the energy and particle density of a trapped Bose gas to leading order of the small diluteness parameter $\bar{\rho}a^3$ (where $\bar{\rho}$ is the mean density) in the limit $N \to \infty$ and $a \to 0$, subject to Na fixed. Next define a reduced one-particle density matrix

$$\gamma(\vec{r}, \vec{r}') = N \int \Psi(\vec{r}, \vec{X})\Psi(\vec{r}', \vec{X})d\vec{X}, \tag{8.9}$$

where $\vec{X} = (\vec{r}_2, \ldots, \vec{r}_N)$ and $d\vec{X} = \prod_{j=2}^{N} d^3 r_j$ and $\Psi(\vec{r}_1, \ldots, \vec{r}_N)$ is the nonnegative, normalized and completely symmetric ground state of H, corresponding to energy $E^{QM}(N, a)$. Then complete (or 100 %) BEC is defined to be the property that $\frac{1}{N}\gamma(\vec{r}, \vec{r}')$ becomes a simple product $f(\vec{r})f(\vec{r}')$, as $N \to \infty$. The function f is called the condensate wave function. It can be shown [15] that in the GP limit, i.e., $N \to \infty$ with $g = 4\pi Na$ fixed, the condensate wave function becomes the GP minimizer ϕ^{GP}

$$\lim_{N \to \infty} \frac{1}{N}\gamma(\vec{r}, \vec{r}') = \phi^{GP}(\vec{r})\phi^{GP}(\vec{r}'). \tag{8.10}$$

This shows that there is complete Bose–Einstein condensation into the state that minimizes the Gross–Pitaevskii energy functional. In fact it can also be shown that there is 100 % condensation for all n-particle reduced density matrices of Ψ, with n fixed as $N \to \infty$. Furthermore, it can be shown [16] that for fixed $g = 4\pi Na$

$$\lim_{N \to \infty} \frac{1}{N}E^{QM}(N, a) = E^{GP}(g),$$

$$\lim_{N \to \infty} \frac{1}{N}\rho(\vec{r}) = |\phi^{GP}(\vec{r})|^2, \tag{8.11}$$

where $\rho(\vec{r}) = \gamma(\vec{r}, \vec{r})$ is the density of the ground state of H.

Other relations between the quantum mechanical solution of H and the solution of the GP equation can be derived and are listed below. Let ϕ_1 denote the scattering solution of v_1, subject to normalization $\lim_{r \to \infty} \phi_1(\vec{r}) = 1$ and $s = \int |\nabla \phi_1|^2/4\pi$. Then $0 < s \le 1$ and the following relations hold [15]

$$\lim_{N \to \infty} \int |\nabla_{\vec{r}_1} \Psi(\vec{r}_1, \vec{X})|^2 d^3 r_1 d\vec{X} = \int |\nabla \phi^{GP}(\vec{r})|^2 d^3 r + gs \int |\phi^{GP}(\vec{r})|^4 d^3 r$$

$$\lim_{N \to \infty} \int V(\vec{r}_1)|\nabla_{\vec{r}_1} \Psi(\vec{r}_1, \vec{X})|^2 d^3 r_1 d\vec{X} = \int V(\vec{r})|\phi^{GP}(\vec{r})|^2 d^3 r$$

$$\lim_{N \to \infty} \frac{1}{2} \sum_{j=2}^{N} \int v(\vec{r}_1 - \vec{r}_j)|\Psi(\vec{r}_1, \vec{X})|^2 d^3 r_1 d\vec{X} = (1 - s)g \int |\phi^{GP}(\vec{r})|^4 d^3 r. \tag{8.12}$$

Additionally, several mathematically rigorous works on BEC and GP theory have been reported since the original pioneering works [17–22]. Mathematical derivation of GP equation of a rotating Bose gas was done by Seiringer and other authors [23–27].

8.3 Many-Body Approach

The ideal theoretical approach would be the solution of the complete many-body Schrödinger equation. However, it is clear from our earlier discussion that an exact solution of this equation is not feasible. In Chap. 4, we saw the galloping complexity of the procedure, as the number of particles increases from three to four. In a laboratory Bose condensate, the number of atoms lies in the range from a few thousand to a few million. It is impossible to handle such a large number exactly. In Chap. 7, we saw that the subset of potential harmonics (PH) of the full hyperspherical harmonics (HH) basis, makes its application quite manageable, if only two-body correlations are retained. The subset keeps only four active degrees of freedom, while freezing the remaining inert degrees of freedom. Indeed correlations higher than two-body one are absent in laboratory condensates (for which $A < 10^7$), except those with very large s-wave scattering length (a_s) produced artificially by Feshbach resonance. Thus the potential harmonics expansion method (PHEM) can be used to expand the condensate wave function [28, 29].

In a not-too-dense condensate, only two-body interactions are relevant, with each pair contributing separately. In a dilute condensate, when the (ij)-pair interacts, remaining particles are merely spectators. Hence correlation arising from the interacting pair only appears in the (ij) Faddeev component. Contributions from all pairs give rise to all two-body correlations in the condensate wave function. To handle this efficiently, we can write the full wave function (ψ) as a sum of Faddeev components (ϕ_{ij}) of all interacting pairs

$$\psi = \sum_{i<j=2}^{A} \phi_{ij}(\vec{r}_1, \ldots, \vec{r}_A). \tag{8.13}$$

In principle, each ϕ_{ij} is a function of position vectors of all particles, if many-body correlations are present in a dense system. But for a dilute condensate, since ϕ_{ij} has only particles i and j correlated, it is a function of $\vec{r}_{ij}$ and does not depend on the individual position vectors of all the remaining $(A - 2)$ particles (for simplicity, we consider a system of spinless identical bosons, each of mass m). The only dependence on these variables will be through a dependence on the hyperradius (r), and we can write (to be consistent with the BEC literature, we use r for hyperradius in this chapter)

$$\psi = \sum_{i<j=2}^{A} \phi_{ij}(\vec{r}_{ij}, r). \tag{8.14}$$

We can then expand $\phi_{ij}(\vec{r}_{ij}, r)$ in the PH subset appropriate for (ij)-pair. The Faddeev component satisfies the Faddeev equation

$$(T + V_{\text{trap}} - E_T)\phi_{ij}(\vec{r}_{ij}, r) = -V(\vec{r}_{ij})\psi, \tag{8.15}$$

where T is the total kinetic energy operator $T = \sum_{i=1}^{A} -\frac{\hbar^2}{2m}\nabla_{\vec{r}_i}^2$, $V_{\text{trap}} = \sum_{i=1}^{A}$
$V_{\text{trap}}(\vec{r}_i) = \sum_{i=1}^{A} \frac{1}{2}m\omega^2 r_i^2$, E_T is the total energy and ψ is given by Eq. (8.14). $V(\vec{r}_{ij})$
is the potential for (ij)-pair interaction. Summing both sides of Eq. (8.15) over all
pairs, i.e., applying $\sum_{i<j=2}^{A}$ on both sides, we get back the original Schrödinger
equation for the A body system, Eq. (4.4). This shows that the Faddeev equation is
equivalent to the Schrödinger equation. In a similar fashion, the Faddeev equation
for the relative motion is

$$(T' + V'_{\text{trap}} - E_R)\Phi_{ij}(\vec{r}_{ij}, r) = -V(\vec{r}_{ij})\Psi, \tag{8.16}$$

where $T' = \sum_{i=1}^{N} -\frac{\hbar^3}{m}\nabla_{\xi_i}^2$ (with $N = A - 1$) is the kinetic energy operator of the
relative motion, $V'_{\text{trap}} = \sum_{i=1}^{N} \frac{1}{2}m\omega^2 \xi_i^2 = \frac{1}{2}m\omega^2 r^2$ and E_R is the relative energy. Ψ
is the total relative wave function and Φ_{ij} is its Faddeev component, so that

$$\Psi = \sum_{i<j=2}^{A} \Phi_{ij}. \tag{8.17}$$

As in Chap. 7, we redefine the Jacobi vectors for each (ij), with $\vec{\xi}_N = \vec{r}_{ij}$. Thus Φ_{ij}
depends on $\vec{\xi}_N$ and r only. Corresponding PH is independent of $\vec{\xi}_1, \ldots, \vec{\xi}_{N-1}$. Hence
$l_1 = l_2 = \cdots = l_{N-1} = 0$, $m_1 = m_2 = \cdots = m_{N-1} = 0$, $n_2 = n_3 = \cdots = n_{N-1} =$
0 (note that $n_1 \equiv 0$) and $L_1 = L_2 = \cdots = L_{N-1} = 0$. All the global quantum numbers become associated with $\vec{\xi}_N$ only and we have $l_N = l$, $m_N = m$, $n_N = K = a$
nonnegative integer and $L_N = 2K + l$, as in Eq. (7.4). Using the notations of Chaps. 4
and 7, the PH is given by Eq. (7.5)

$$\mathcal{P}_{2K+l}^{l,m}(\Omega_N^{(ij)}) = Y_{l,m}(\vartheta_N, \varphi_N) \, {}^{(N)}\mathcal{P}_{2K+l}^{l,0}(\phi_N)\mathcal{Y}_{[0]}(3N - 3), \tag{8.18}$$

where $\mathcal{Y}_{[0]}(3N - 3)$ is the HH of order zero in $(3N - 3)$-dimensional space,
given by Eq. (7.6). A hyperradius in the $(3N - 3)$-dimensional space spanned by
$\{\vec{\xi}_1, \ldots, \vec{\xi}_{N-1}\}$ is defined as $\rho_{ij} = \sum_{i=1}^{N-1} \xi_i^2$. Then the hyperangle ϕ_N is defined
through

$$r_{ij} = r \cos\phi_N \qquad\qquad \rho_{ij} = r \sin\phi_N, \tag{8.19}$$

and (ϑ_N, φ_N) are the polar angles of $\vec{\xi}_N$. Expanding $\Phi_{ij}(\vec{r}_{ij}, r)$ in the subset of PH
given by Eq. (8.18), we have

$$\Phi_{ij}(\vec{r}_{ij}, r) = r^{-(3N-1)/2} \sum_{K'} \mathcal{P}_{2K'+l}^{lm}(\Omega_N^{(ij)})u_{K'}^{l}(r). \tag{8.20}$$

Here $u_K^l(r)$ is the hyperspherical partial wave. The factor $r^{-(3N-1)/2}$ is included
to remove first derivatives with respect to r. To obtain a set of coupled differential equation (CDE), we substitute Eq. (8.20) in Faddeev equation (8.16), multiply

by $\mathcal{P}_{2K+l}^{lm}{}^{*}(\Omega_N^{(ij)})$, integrate over $d\Omega_N^{(ij)}$, and use orthonormalization relation of PH functions (Eq. (7.7)) to get

$$
\left[-\frac{\hbar^2}{m} \frac{d^2}{dr^2} + \frac{1}{2}m\omega^2 r^2 + \frac{\hbar^2}{m} \frac{\mathcal{L}_K(\mathcal{L}_K + 1)}{r^2} - E_R \right] u_K^l(r)
$$
$$
+ \sum_{K'} f_{K'l}^2 V_{KK'}(r) u_{K'}^l(r) = 0,
\tag{8.21}
$$

where E_R is the energy of relative motion and

$$
\mathcal{L}_K = 2K + l + \frac{3N - 3}{2}
$$

$$
f_{Kl}^2 = \sum_{p<q=2}^{A} \langle \mathcal{P}_{2K+l}^{lm}(\Omega_N^{(ij)}) | \mathcal{P}_{2K+l}^{lm}(\Omega_N^{(pq)}) \rangle
$$

$$
= \sum_{p<q=2}^{A} \int \mathcal{P}_{2K+l}^{lm}{}^{*}(\Omega_N^{(ij)}) \, \mathcal{P}_{2K+l}^{lm}(\Omega_N^{(pq)}) \, d\Omega_N^{(ij)}
$$

and

$$
V_{KK'}(r) = \int \mathcal{P}_{2K+l}^{lm}{}^{*}(\Omega_N^{(ij)}) \, V(r_{ij}) \, \mathcal{P}_{2K'+l}^{lm}(\Omega_N^{(ij)}) \, d\Omega_N^{(ij)}
$$

$$
= (h_K^{\alpha\beta} h_{K'}^{\alpha\beta})^{\frac{1}{2}} \int_{-1}^{1} P_K^{\alpha\beta}(z) V\left(r\sqrt{\frac{1+z}{2}}\right) P_{K'}^{\alpha\beta}(z) W_l(z) dz,
\tag{8.22}
$$

with $\alpha = \frac{3N-5}{2}$ and $\beta = l + \frac{1}{2}$ and the weight function of Jacobi polynomial $W_l(z) = (1-z)^\alpha (1+z)^\beta$. The overlap f_{Kl}^2 is given by Eq. (7.23). Note that the particle label (ij) is arbitrary and for convenience can be chosen as (12).

A numerical solution of Eq. (8.21) encounters a serious problem in computing $V_{KK'}(r)$. From the third of Eq. (8.22), we see that the integrand involves the weight function $W_l(z) = (1-z)^\alpha (1+z)^\beta$, which for large A, is extremely strongly peaked, within an extremely narrow interval near $z = -1$. As a result, any standard quadrature gives nearly zero for the integral. This problem can be solved by evaluating the integral in gradually increasing subintervals, starting from an extremely small first subinterval (see Chap. 10, Sect. 10.3).

Symmetry of the Wave Function

As we have a system of bosons, the wave function must be symmetric under any pair exchange. Since Eq. (8.17) shows that the condensate relative wave function Ψ is a sum over all Faddeev components, it is symmetric under any pair exchange, provided Φ_{ij} is symmetric under $i \longleftrightarrow j$. The latter is easily achieved by taking l even. However, under P_{ik}, where $k \neq i$ or j, recoupling of angular momenta complicates the picture. The situation becomes much simpler, for states with zero total orbital angular momentum ($l = 0$). In this case, angular momenta of all intermediate re-couplings become zero and the $l = 0$ states (including the ground state) become totally symmetric under any pair exchange. Indeed the use of PH subset assumes that each spectator contributes zero angular momentum. Hence the state will be strictly symmetric under any pair exchange, only if the interacting pair also contributes zero angular momentum (i.e., $l_1 = l_2 = \cdots = l_A = 0$ and $\vec{l} = \vec{l}_1 + \cdots + \vec{l}_A = 0$). States with $l > 0$ need a special treatment. The present procedure will make states, with even values of $l > 0$, only approximately symmetric under any pair exchange.

8.4 Need for a Short-Range Correlation Function

The procedure mentioned in the previous section faces two problems:

1. If a realistic interatomic interaction is used for $V(r_{ij})$ in Eq. (8.22), then the solution of the CDE, Eq. (8.21), gives an unphysically strongly bound ground state with large negative energy, even for small positive a_s. The correct energy per particle should be slightly larger than the ground state energy of the pure harmonic oscillator.
2. Also the convergence rate of expansion (8.20) turns out to be very slow.

The reason for the first problem is that a realistic interatomic interaction is attractive for $r_{ij} \gtrsim 5\ a_0$ and is too large negative in BEC energy scale. From the last of Eq. (8.22), we see that the integrand in the region $z \to -1$ (i.e., $r_{ij} \to 0$) contributes the largest (since for large A, $W_l(z)$ is very strongly peaked near $z = -1$) to the integral in $V_{KK'}(r)$ for any r. Its contribution becomes large negative and a solution of Eq. (8.21) corresponds to a molecular bound state. For $a_s > 0$, effective potential has to be repulsive. How can we get an *effective repulsion* from a strongly attractive $V(r_{ij})$? The answer lies in the *extremely low energy scale* of the BEC. We saw in Sect. 8.1 that at such near-zero energy the effective interaction is governed by the s-wave scattering length a_s and in the lowest approximation it becomes a contact potential whose strength is proportional to a_s. But in Sect. 8.2, we discussed the inadequacy of a pure contact interaction.

The relative wave function $\eta(r_{ij})$ of the pair of interacting bosons at near-zero energy very closely satisfies the zero-energy two-body Schrödinger equation

$$\left[-\frac{\hbar^2}{m} \frac{1}{r_{ij}^2} \frac{d}{dr_{ij}} \left(r_{ij}^2 \frac{d}{dr_{ij}} \right) + V(r_{ij}) \right] \eta(r_{ij}) = 0. \tag{8.23}$$

Now, in the $r_{ij} \to 0$ limit, $\eta(r_{ij})$ must vanish, since $V(r_{ij})$ has a strong short-range repulsion. In the $r_{ij} \to \infty$ limit, where the potential vanishes, the two-body wave function quickly (in BEC length scale) attains the form [4]

$$\eta(r_{ij}) = C \left(1 - \frac{a_s}{r_{ij}} \right). \tag{8.24}$$

Fig. 8.1 Plot of $\eta(r_{ij})$ against r_{ij} (in units of a_{ho}) in log scale for the ^{87}Rb atoms having $a_s = 100\ a_0$. Since the rapid change in $\eta(r_{ij})$ from negative to positive values takes place in a very narrow region, its detailed structure for small r_{ij} has been displayed in the semilog plot. Value of the asymptotic normalization is taken as $C = 1$

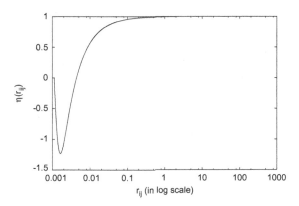

Figure 8.1 is a plot of $\eta(r_{ij})$ for the ^{87}Rb atoms interacting through the van der Waals potential for $a_s = 100\ a_0\ (= 0.00433\ a_{\text{ho}})$. We see that $\eta(r_{ij})$ attains its asymptotic value very quickly. Rapid change of $\eta(r_{ij})$ occurs within about $0.01\ a_{\text{ho}}$. The detailed structure of $\eta(r_{ij})$ for small r_{ij} is displayed in the log scale for r_{ij}. We see that the particles have a very small probability to come close to each other. But expansion (8.20) shows that for $r_{ij} \to 0$ (i.e., for $\phi_N \to \frac{\pi}{2}$), $\Phi_{ij}(\vec{r}_{ij}, r)$ remains finite for $K = 0$ (for $K = 0$ the PH is a constant, see Eqs. (8.18), (8.19) and (4.15)). In order that Φ_{ij} may vanish in this limit, the sum in Eq. (8.20) must contain a large number of K values. This explains the slow rate of convergence (problem (2) mentioned above). Clearly, the small r_{ij} behavior of Φ_{ij} is given by $\eta(r_{ij})$. Thus we can take $\eta(r_{ij})$ as a *short-range correlation function* and replace Eq. (8.20) by [29, 30]

$$\Phi_{ij}(\vec{r}_{ij}, r) = r^{-(3N-1)/2} \sum_{K'} \mathcal{P}^{lm}_{2K'+l}(\Omega_N^{(ij)}) \eta(r_{ij}) u^l_{K'}(r). \qquad (8.25)$$

The basis $\{\mathcal{P}^{lm}_{2K+l}(\Omega_N^{(ij)}) \eta(r_{ij})\}$ is called the *correlated potential harmonic* (CPH) basis and its use is called the CPH expansion method (CPHEM). This basis is non-orthogonal and can be handled by a standard procedure. But in general, it will lead to a lot of numerical difficulties. Substitution of Eq. (8.25) in Eq. (8.16) and projection on a particular PH, $\mathcal{P}^{lm}_{2K+l}(\Omega_N^{(ij)})$, introduces the overlap matrix $A_{KK'}(r) \equiv \langle \mathcal{P}^{lm}_{2K+l}(\Omega_N^{(ij)}) | \mathcal{P}^{lm}_{2K'+l}(\Omega_N^{(ij)})\ \eta(r_{ij}) \rangle$. Then its first and second hyperradial derivatives appear in the CDE and the numerical procedure to handle the non-orthogonal, r-dependent basis becomes very slow. However, a great deal of simplification is possible in an approximate treatment valid for a trapped dilute BEC. A numerical solution of Eq. (8.23) shows that $\eta(r_{ij})$ is appreciably different from its asymptotic value C (independent of r) only in a tiny interval of small values of r_{ij} (Fig. 8.1). Thus, except for small values of r, the overlap $A_{KK'}(r)$ is nearly independent of r. Since $\eta(r_{ij})$ is practically independent of r_{ij}, it is also seen that $A_{KK'}(r)$ is approximately proportional to $\delta_{KK'}$. Now the trapped condensate in the hyperradial space resides around the minimum of the effective potential, which occurs at approximately $r_{\min} = \sqrt{3A}\ a_{\text{ho}}$. Thus, for large A and for values of r close to $r_{\min}$, the overlap matrix can be approximated by $A_{KK'}(r) \approx C'\delta_{KK'}$, with C' a constant independent of r. With this assumption, the CDE simplifies greatly, as we see below.

Substituting Eq. (8.25) in Eq. (8.16), projecting on the PH $\mathcal{P}^{lm}_{2K+l}(\Omega^{(ij)}_N)$ and using Eqs. (7.20) and (7.22) we have (note that the indices (ij) is arbitrary and can be replaced by (12))

$$\left[-\frac{\hbar^2}{m}\frac{d^2}{dr^2} + \frac{1}{2}m\omega^2 r^2 + \frac{\hbar^2}{m}\frac{\mathcal{L}_K(\mathcal{L}_K+1)}{r^2} - E_R\right]u^l_K(r)$$
$$+ \sum_{K'} f^2_{K'l}\tilde{V}_{KK'}(r)u^l_{K'}(r) = 0, \tag{8.26}$$

where

$$\tilde{V}_{KK'}(r) = \int \mathcal{P}^{lm}_{2K+l}{}^*(\Omega^{(12)}_N)\, V(r_{12})\, \mathcal{P}^{lm}_{2K+l}(\Omega^{(12)}_N)\, \eta(r_{12})\, d\Omega^{(12)}_N$$
$$= (h^{\alpha\beta}_K h^{\alpha\beta}_{K'})^{-\frac{1}{2}} \int_{-1}^{1} P^{\alpha\beta}_K(z) V\left(r\sqrt{\frac{1+z}{2}}\right) \eta\left(r\sqrt{\frac{1+z}{2}}\right) P^{\alpha\beta}_{K'}(z) W_l(z) dz. \tag{8.27}$$

The quantity f^2_{Kl} is given by Eq. (7.23). The asymptotic value of $\eta(r_{ij})$ is given by Eq. (8.24), where $C = 1/C'$. Value of C' is proportional to the initially chosen arbitrary asymptotic value (say, 1) of η. Hence η whose asymptotic value is C is independent of an arbitrary normalization. As a simple procedure, instead of evaluating C by computing the overlap, one can treat it as an empirical parameter, whose value is determined by fitting with some known property. Note that C is independent of A.

Equation (8.26) can be put in a more convenient (symmetric) form, by multiplying through by the constant f_{Kl}

$$\left[-\frac{\hbar^2}{m}\frac{d^2}{dr^2} + \frac{1}{2}m\omega^2 r^2 + \frac{\hbar^2}{m}\frac{\mathcal{L}_K(\mathcal{L}_K+1)}{r^2} - E_R\right]U_{Kl}(r)$$
$$+ \sum_{K'} \overline{V}_{KK'}(r)U_{K'l}(r) = 0, \tag{8.28}$$

where

$$U_{Kl}(r) = u^l_K(r)f_{Kl}$$
$$\overline{V}_{KK'}(r) = f_{Kl}\tilde{V}_{KK'}(r)f_{K'l}. \tag{8.29}$$

Computation of the potential matrix element $\overline{V}_{KK'}(r)$ and solution of the CDE (8.28) are straight forward. However numerical difficulties appear for large A, which can be handled using Eqs. (9.37) and (9.38), see Ref. [31]. For bound states, Eq. (8.28) is to be solved subject to appropriate boundary conditions, viz., $U_{Kl}(r)$ vanishes as $r \to 0$ and $r \to \infty$.

8.5 Results for Repulsive and Attractive Condensates

Initially, the PHEM without the short-range correlation function (SRCF) was applied to typical BECs: repulsive [28, 32] as well as attractive [33]. However, the difficulty was that it did not give correct results for $A > 50$. Then it was realized that the SRCF is an essential ingredient in the PHEM. In the following, we describe a brief history of such calculations to explain original experiments: ^{87}Rb condensate [34] in the trap used at Joint Institute for Laboratory Astrophysics (JILA), as a typical repulsive condensate and ^{7}Li [35] and ^{85}Rb [36] condensates as examples of attractive condensates.

For the ^{87}Rb condensate the natural s-wave scattering length (a_s) is $100\ a_0$. For a simple application a spherical trap of frequency $\omega = 77.78$ Hz (which is the geometric average value for an axially symmetric trap used in the original experiment at JILA) is chosen. This value of ω corresponds to $a_{ho} = 23095\ a_0$. Thus we see that the range of actual interatomic interaction ($\sim 20\ a_0$) is quite small compared to $|a_s|$, which, in turn, is much less than a_{ho}. Thus the conditions for the successful application of the PHEM are satisfied. For the interatomic potential, a commonly chosen potential is the van der Waals potential with a hard core of radius r_c

$$
\begin{aligned}
V(r_{ij}) &= \infty && \text{for } r_{ij} < r_c \\
&= -\frac{C_6}{r_{ij}^6} && \text{for } r_{ij} \geq r_c.
\end{aligned}
\tag{8.30}
$$

The van der Waals potential is a realistic interatomic potential with known value of C_6 for a given type of atoms [4]. The exact mathematical dependence of the very short-range repulsion is not known experimentally. It is modeled by the hard core repulsion of Eq. (8.30). Value of r_c is adjusted, so that the solution $\eta(r_{ij})$ of the two-body equation (8.23) has an asymptotic form given by Eq. (8.24) with the appropriate value of a_s [4]. For a selected value of r_c, value of a_s is obtained from the asymptotic region of $\eta(r_{ij})$ subject to the boundary condition $\eta(r_c) = 0$. A plot of a_s against r_c for the van der Waals potential, Eq. (8.30). is presented in Fig. 8.2. One can see that there are an infinite number of discontinuous branches. The right most branch corresponds to zero nodes in $\eta(r_{ij})$. As r_c is decreased from a large value, a_s starts from a positive value, then decreases slowly, crosses zero and then rapidly decreases toward $-\infty$ at a particular $r_c = r_{c1}$. As r_c is decreased infinitesimally from r_{c1}, value of a_s jumps discontinuously and starts decreasing again from $+\infty$, as r_c is further decreased. Finally, at another particular value $r_c = r_{c2}$, value of a_s passes through another infinite discontinuity. This second branch corresponds to one node in $\eta(r_{ij})$. As r_c is decreased further, this pattern is repeated, with the appearance of one extra node in $\eta(r_{ij})$ for each new branch. The widths of successive branches reduce rapidly, to accommodate an infinite number of branches at $r_c = 0$. Appearance of an extra node in $\eta(r_{ij})$ corresponds to the appearance of an extra virtual state of the two-body system.

Fig. 8.2 Plot of a_s against r_c (both in units of a_{ho}) for Rb atoms

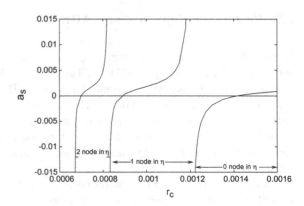

For the numerical calculation with a negative a_s, the negative region of the first branch is chosen. On the other hand, for a positive a_s, the positive region of the second branch (with one node in $\eta(r_{ij})$) is chosen. We saw a plot of $\eta(r_{ij})$ for a positive a_s (=100 a_0 = 0.00433 a_{ho}) in Fig. 8.1, from which we see that it starts off with a negative value from $r_{ij} = r_c$, has a node and then becomes positive, approaching its asymptotic value in a short interval of r_{ij} As an example of the negative a_s, a plot of $\eta(r_{ij})$ for $a_s = -1.836 \times 10^{-4}$ a_{ho} is displayed in Fig. 8.3. It has no nodes and is purely positive. Comparing Eq. (8.27) with Eq. (8.22), we see that the effective two-body potential for the dilute BEC becomes

$$V_{\text{eff}}(r_{ij}) = V(r_{ij})\eta(r_{ij}). \qquad (8.31)$$

Since $V(r_{ij})$ is negative for all $r_{ij} > r_c$, we see that $V_{\text{eff}}(r_{ij})$ is also purely negative for $a_s < 0$, since corresponding $\eta(r_{ij})$ is purely positive. However, $V_{\text{eff}}(r_{ij})$ is less attractive than $V(r_{ij})$. We discussed earlier that it should be so. On the other hand, for $a_s > 0$, $\eta(r_{ij})$ is initially negative and then it becomes positive. Hence in this case,

Fig. 8.3 Plot of $\eta(r_{ij})$ against r_{ij} (in units of a_{ho}) in log scale for the ^{85}Rb atoms having $a_s = -1.836 \times 10^{-4}$ a_{ho}. This has no nodes, but increases rapidly from zero at $r_{ij} = r_c$ to its asymptotic value in a tiny interval. Note the log scale for r_{ij}. Value of asymptotic normalization is taken as $C = 1$

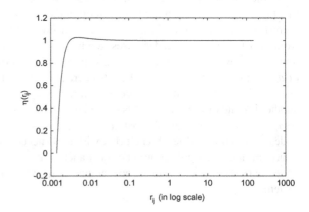

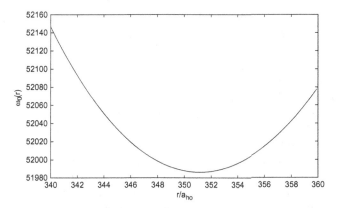

Fig. 8.4 Effective hyperradial potential $w_0(r)$ against r in units of a_{ho} for 10^4 atoms of ^{87}Rb, having $a_s = 4.33 \times 10^{-3}\ a_{ho}$ in a harmonic trap. The effective potential for a repulsive condensate has an absolute minimum, where the stable condensate resides

$V_{eff}(r_{ij})$ has a short-range repulsive part just outside the hard core and then a reduced attractive part. The net effect of this is a repulsive effective two-body interaction. The effective potential (in harmonic oscillator units, o.u.) in the hyperradial space, $w_0(r)$ (see Sect. 10.2.3 of Chap. 10) for 10^4 atoms of ^{87}Rb with a positive a_s ($=0.00433\ a_{ho}$) in a harmonic trap is shown in Fig. 8.4. This has a minimum in a stable region. Thus the repulsive condensate will be stable, as expected.

The effective potential (in o.u.) for an attractive condensate of ^{85}Rb atoms having $a_s = -1.836 \times 10^{-4}\ a_{ho}$ in a harmonic trap is shown in Fig. 8.5. The stable well of the positive a_s is replaced for an attractive condensate by a shallow well and a deep attractive well (called the *collapse region*) at a smaller r, which are separated by an intermediate barrier. Hence the condensate residing in the shallow well has a

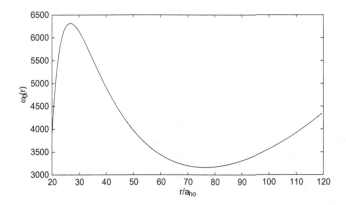

Fig. 8.5 Plot of effective hyperradial potential $w_0(r)$ against r/a_{ho} for 2400 atoms of ^{85}Rb, having $a_s = -1.836 \times 10^{-4}\ a_{ho}$ in a harmonic trap

probability of transmission through the intermediate barrier into the deep well, where cluster states are formed. Thus the condensate has a finite life time and eventually forms clusters. Consequently, the shallow well is referred to as the *metastable region*. This agrees with the experimental observation [36]. With increase in A, the stable well of Fig. 8.4 has steeper walls and the position of the minimum shifts to the right. The changes are similar, when a_s is increased keeping A the same. Thus, with increase of either A or a_s, the condensate remains stable. On the other hand, for the attractive condensate, either increasing $|a_s|$ (i.e., making a_s more negative), or increasing A, the metastable well becomes shallower and the intermediate barrier becomes lower. Hence the probability of transmission into the collapse region increases. At a particular value of A ($=A_{cr}$) for a fixed a_s, the minimum of metastable region merges with the maximum of the intermediate barrier (resulting in a point of inflection in $\omega_0(r)$). For $A > A_{cr}$, there is no metastable region and the attractive condensate ceases to exist. This is called the collapse of the attractive condensate and A_{cr} is called the critical number. Critical numbers calculated by CPHEM technique [30] agree nicely with experimental values [36].

Various aspects of the BEC have been investigated and reported in a large number of publications since the 1980s. Most of the theoretical works are based on the GP equation. We will not discuss them, but mention only a few in connection with the calculations using CPHEM. The CHPEM technique has been used in BEC theory since the beginning of the last decade by a number of groups around the world. Many calculations using CPHEM have been reported by Das, Chakrabarti, Canuto, Sofianos, Salasnich and their collaborators, which will be mentioned below. Besides these, the essentially exact diffusion Monte Carlo (DMC) method by Blume and collaborators has used hyperspherical description for BEC [37]. Adiabatic hyperspherical approach has also been used in the study of trapped bosons by Sorensen et al. [38].

The lowest lying state in the stable or metastable well (for the repulsive or attractive condensate respectively) is the ground state of the condensate. Excitations in these wells produce excited states of the condensate. Properties of such states and the condensate as a whole have been calculated [39–41], which compare favorably with experimental observations. Since the GP equation is based on a mean field theory and ignores correlations, the effects beyond mean field using CPHEM were investigated and found to be appreciable [31, 42]. Also the shape independence of the two-body potential, which is inherent in the GP theory, was tested by the CPHEM technique and found to be valid for dilute BEC containing a small number of bosons [43], but it is violated for large A or strongly interacting condensates [44]. Strongly interacting condensates are generated by large s-wave scattering length (a_s). Such condensates have been achieved in the laboratory by Feshbach resonance. This has generated a lot of interest in strongly interacting Bose gas with a large a_s. The behavior of a trapped dilute Bose gas with a large scattering length has been investigated in Ref. [44]. GP equation is commonly used to investigate ultracold Bose gases [3, 45]. Ground state properties of ultracold Bose gases at large scattering length were calculated using the GP equation [46]. Ground state properties at zero temperature of an attractive BEC was obtained by the correlated potential harmonic method [47].

Thermodynamic quantities can be calculated using a large number of excited levels, which are populated at a given temperature according to Bose distribution law [48]. Condensate fraction ($\frac{A_0(T)}{A}$) is defined as the fraction of atoms in the ground state at a temperature T. At the critical temperature (T_c), the condensate fraction suddenly becomes microscopically small. Critical temperature, condensate fraction, heat capacity, etc. calculated by the CPHEM technique agree with experimental results and other calculations [48–50]. Effect of interaction on the thermodynamics of a repulsive BEC was studied in Ref. [51]. Corresponding calculations based on the GP equation can be found in Ref. [52].

Besides the general properties of BEC, the CPHEM was used to investigate the stability of the attractive bosonic cloud [53] and its destruction under tightening of the trap [54]. This method has also been adopted in studies of anharmonic traps [55], resonance and quantum tunneling [56], information entropy [57], etc. These calculations demonstrate that the CPHEM technique is well suited for the Bose–Einstein condensates.

References

1. Bose, S.N.: Z. Phys. **26**, 178 (1924)
2. Einstein, A., Sitzber. Kgl. Preuss. Akad. Wiss. p. 261 (1924); *ibid* p. 3 (1925)
3. Dalfovo, F., Giorgini, S., Pitaevskii, L.P., Stringari, S.: Rev. Mod. Phys. **71**, 463 (1999)
4. Pethick, C.J., Smith, H.: Bose–Einstein Condensation in Dilute Gases. Cambridge University Press, Cambridge (2002)
5. Pitaevskii, L., Stringari, S.: Bose–Einstein Condensation. Oxford University Press, Oxford (2003); Dalfovo, F., Pitaevskii, L., Stringari, S.: Encyclopedia of Mathematical Physics, vol. 1, p. 312. Elsevier, Amsterdam (2006)
6. Albeverio, S., Gesztesy, F., Høegh-Krohn, R., Holden, H.: Solvable Models in Quantum Mechanics. Springer, Berlin (1980)
7. Correggi, M., Dell'Antonio, G., Finco, D., Michelangeli, A., Teta, A.: A class of Hamiltonians for a three-particle fermionic system at unitarity. arXiv:1505.04132v1 [math-ph] 15 May 2015
8. Nam, P.T., Rougerie, N., Seiringer, R.: Ground states of large Bose systems: the Gross–Pitaevskii limit revisited. arXiv:1503.07061v2 [math-ph] 7 April 2015
9. Tiwari, R.P., Shukla, A.: Comput. Phys. Commun. **174**, 966 (2006)
10. Jastrow, R.: Phys. Rev. **98**, 1479 (1955)
11. Yngvason, J.: Topics in the mathematical physics of cold Bose gases, presented at the 5th Warsaw school of statistical physics in June 2013. arXiv:1402.0706v2 [cond-mat.quant-gas] 14 Feb 2014
12. Seiringer, R.: Cold quantum gases and Bose–Einstein condensation. In: Guiliani, A., Mastropietro, V., Yngvason, J. (eds.) Quantum Many Body Systems. Lecture Notes in Mathematics, vol. 2051, p. 5592. Springer, New York (2012)
13. Seiringer, R.: Dilute trapped Bose gases and Bose–Einstein condensation. In: Derezinski, J., Siedentop, H. (eds.) Large Coulomb Systems. Lecture Notes in Physics, vol. 695, pp. 251–276. Springer, New York (2006)
14. Lieb, E.H., Seiringer, R., Solovej, J.P., Yngvason, J.: The mathematics of Bose gas and its condensation. arXiv:0610117v1 [cond-mat.stat-mech] 4 Oct 2006. (Originally published by Birkhüser Verlag (Basel-Boston-Berlin) in 2005 under the title The Mathematics of the Bose Gas and its Condensation as number 34 of its Oberwolfach Seminar series)
15. Lieb, E.H., Seiringer, R.: Phys. Rev. Lett. **88**, 170409 (2002)

16. Lieb, E.H., Seiringer, R., Yngvason, J.: Phys. Rev. A **61**, 043602 (2000)
17. Lieb, E.H., Yngvason, J.: Phys. Rev. Lett. **80**, 2504 (1998)
18. Lieb, E.H., Seiringer, R., Yngvason, J.: Phys. Rev. Lett. **94**, 080401 (2005)
19. Erdös, L., Schlein, B., Yau, H.-T.: Invent. Math. **167**, 515 (2007); Phys. Rev. A **78**, 053627 (2008)
20. Erdös, L., Schlein, B., Yau, H.-T.: J. Am. Math. Soc. **22**, 1099 (2009); Ann. Math. **172**(2), 291 (2010)
21. Lewin, M., Nam, P.T., Rougerie, N.: Adv. Math. **254**, 570 (2014); Trans. Am. Math. Soc. in press. arXiv:1405.3220 (2014)
22. Pickl, P.: Rev. Math. Phys. **27**, 1550003 (2015)
23. Seiringer, R.: Commun. Math. Phys. **229**, 491 (2002)
24. Seiringer, R.: J. Phys. A **36**, 9755 (2003)
25. Lieb, E.H., Seiringer, R.: Commun. Math. Phys. **264**, 505 (2006)
26. Lieb, E.H., Seiringer, R., Yngvason, J.: Phys. Rev. A **79**, 063626 (2009)
27. Lewin, M., Seiringer, R.: J. Stat. Phys. **137**, 1040 (2009)
28. Das, T.K., Chakrabarti, B.: Phys. Rev. A **70**, 063601 (2004)
29. Das, T.K., Canuto, S., Kundu, A., Chakrabarti, B.: Phys. Rev. A **75**, 042705 (2007)
30. Das, T.K., Kundu, A., Canuto, S., Chakrabarti, B.: Phys. Lett. A **373**, 258 (2009)
31. Sofianos, S.A., et al.: Phys. Rev. A **87**, 013608 (2013)
32. Chakrabarti, B., Kundu, A., Das, T.K.: J. Phys. B **38**, 2457 (2005)
33. Kundu, A., Chakrabarti, B., Das, T.K., Canuto, S.: J. Phys. B **40**, 2225 (2007)
34. Anderson, M.H., Ensher, J.R., Matthews, M.R., Wieman, C.E., Cornell, E.A.: Science **269**, 198 (1995)
35. Bradley, C.C., Sackett, C.A., Tollett, J.J., Hulet, R.G.: Phys. Rev. Lett. **75**, 1687 (1995)
36. Roberts, J.L., et al.: Phys. Rev. Lett. **86**, 4211 (2001)
37. Blume, D., Greene, C.H.: Phys. Rev. A **63**, 063601 (2001); Bortolotti, D.C.E., et al.: Phys. Rev. Lett. **97**, 160402 (2006); Kalas, R.M., Blume, D.: Phys. Rev. A **77**, 032703 (2008)
38. Sorensen, O., et al.: Phys. Rev. A **65**, 051601 (2002); Sorensen, O., Federov, D.F., Jensen, A.S.: Phys. Rev. **66**, 032507 (2002); Phys. Rev. Lett. **89**, 173002 (2002)
39. Biswas, A., Das, T.K.: J. Phys. B **41**, 231001 (2008)
40. Biswas, A., Chakrabarti, B., Das, T.K.: J. Chem. Phys. **133**, 104502 (2010)
41. Sofianos, S.A., et al.: Few-Body Syst. **54**, 1529 (2013)
42. Debnath, P.K., Chakrabarti, B., Das, T.K.: Int. J. Quantum Chem. **111**, 1333 (2011)
43. Chakrabarti, B., Das, T.K.: Phys. Rev. A **78**, 063608 (2008)
44. Lekala, M.L., et al.: Phys. Rev. A **89**, 023624 (2014)
45. Stringari, S.: Phys. Lett. A **347**, 150 (2005)
46. Purwanto, W., Zhang, S.: Phys. Rev. A **72**, 053610 (2005); Zezyulin, D.A., et al.: Phys. Rev. A **78**, 013606 (2008); Song, J.L., Zhou, F.: Phys. Rev. Lett. **103**, 025302 (2009)
47. Chakrabarti, B., Das, T.K., Debnath, P.K.: J. Low Temp. Phys. **157**, 527 (2009)
48. Goswami, S., Das, T.K., Biswas, A.: Phys. Rev. A **84**, 053617 (2011)
49. Goswami, S., Das, T.K., Biswas, A.: J. Low Temp. Phys. **172**, 184 (2013)
50. Haldar, S.K., et al.: Eur. Phys. J. D **68**, 262 (2014)
51. Bhattacharyya, S., Das, T.K., Chakrabarti, B.: Phys. Rev. A **88**, 053614 (2013)
52. Kling, S., Pelster, A.: Phys. Rev. A **76**, 023609 (2007); Dorfman, K.E., et al.:Phys. Rev. A **83**, 033609 (2011)
53. Biswas, A., Das, T.K., Salasnich, L., Chakrabarti, B.: Phys. Rev. A **82**, 043607 (2010)
54. Biswas, A., Chakrabarti, B., Das, T.K., Salasnich, L.: Phys. Rev. A **84**, 043631 (2011)
55. Chakrabarti, B., Das, T.K., Debnath, P.K.: Phys. Rev. A **79**, 053629 (2009)
56. Haldar, S.K., Chakrabarti, B., Das, T.K.: Phys. Rev. A **82**, 043616 (2010)
57. Haldar, S.K., Chakrabarti, B., Das, T.K., Biswas, A.: Phys. Rev. A **88**, 033603 (2013)

Chapter 9
Integro-Differential Equation

Abstract Solving for the complete many-body wave function (instead of partial waves in a PH expansion), one gets an integro-differential equation (IDE). The IDE is derived from PH expansion method. Hence, IDE and PHEM are equivalent. Still IDE has certain advantages: its structure and complexity do not increase with the number of particles. Also, since there is no sum over K, there is no problem of convergence. However, calculation of the kernel function is tricky. Application of IDE to nuclear systems and BEC is discussed.

The potential harmonics (PH) basis is suitable for systems in which correlations higher than two-body correlations can be ignored. We saw in Chap. 7 that the expansion of the wave function in the PH basis and projection on a particular PH give rise to a system of coupled differential equations (CDE). The method, being variational in nature with respect to inclusion of higher harmonics, obeys the Ritz principle. In Chap. 8, we saw that the PH expansion, together with a short-range correlation function, converges very fast for the Bose–Einstein condensate. However, the rate of convergence depends on the nature of the system and may not be desirably fast enough. In fact the PH expansion cannot describe asymptotic part of the many-body wave function exactly, since the expansion must be truncated for a numerical calculation, while higher K-partial waves are strongly pushed out in hyperradial space due to the hyper-centrifugal repulsion which increases rapidly with K. Convergence of energy can take place at a smaller upper limit in K, for which the asymptotic part of wave function may not still be converged. Thus the usefulness of the method depends on the nature of the system specifying how fast the PH expansion converges, which in turn depends on the interparticle interactions. For example, if the sum of pair-wise interactions deviates from a hypercentral form by a small amount, the PH expansion will converge quickly. The success of the potential harmonics expansion method (PHEM) for trapped dilute BEC is due to the fact that the effective interparticle interaction is quite weak compared with the hypercentral trapping potential.

One can argue that it will be more desirable if one solves an equation for the unknown many-body wave function, which is complete, i.e., its expansion in PH basis includes all PH partial waves. This will then lead to an integro-differential equation like the Faddeev or Faddeev–Yakubovsky equations. Indeed, one can develop

© Springer India 2016

T.K. Das, *Hyperspherical Harmonics Expansion Techniques*,
Theoretical and Mathematical Physics, DOI 10.1007/978-81-322-2361-0_9

an integro-differential equation (IDE) starting from the PH expansion, as done originally by Fabre [1]. Later this equation has been rederived and used by several authors
[2–8]. In this chapter, we discuss such a technique. Although the IDE is equivalent to
the PHEM when all partial waves are included, it does not satisfy the Ritz principle.
Hence, there is no control over the precision, which is essentially due to numerical errors. The IDE is exact with respect to inclusion of all partial waves (within
the assumption of two-body correlations only). Moreover, its structure remains the
same and its complexity does not increase with the number of particles. The merit
of the technique depends on an accurate computation of the kernel function. It also
depends on how fast and accurately the integro-differential equation can be solved
numerically. It is formally equivalent to the Faddeev (for three-body systems) or
Faddeev–Yakubovsky (for systems with more than three particles) equations involving two-body correlations only [1]. In the following section, we discuss how the
integro-differential equations can be derived from the PHEM.

9.1 Derivation of IDE

Since the IDE is derived from the PHEM, it also includes two-body correlations,
while effects of higher-body correlations are disregarded. Thus the IDE is applicable
to systems where the PHEM is also applicable, *viz.* in very dilute systems in which
three-body and higher-body collisions can be disregarded. The potential harmonics
basis was introduced in Chap. 7 and its application to Bose–Einstein condensates was
discussed in Chap. 8. We will follow the treatment by Fabre [1] for the derivation of
the IDE, using the notations of Chaps. 7 and 8.

For a dilute system interacting through two-body forces only, the A-body wave
function can be decomposed into Faddeev components

$$\psi(\vec{r}_1, \ldots, \vec{r}_A) = \sum_{i,j>i}^{A} \phi_{ij}(\vec{r}_1, \ldots, \vec{r}_A), \tag{9.1}$$

where the Faddeev component $\phi_{ij}(\vec{r}_1, \ldots, \vec{r}_A)$ corresponds to the ij-pair interacting,
all other particles being inert spectators. If, in addition, the system is sufficiently
dilute, so that only two-body correlations are important, then the ij-Faddeev compo-
nent is a function only of the interacting-pair separation $\vec{r}_{ij} = \vec{r}_j - \vec{r}_i$ and the global
length (hyperradius) ξ, besides the CM coordinate $\vec{R}$

$$\phi_{ij}(\vec{r}_1, \ldots, \vec{r}_A) = \phi_{ij}(\vec{R}, \vec{r}_{ij}, \xi). \tag{9.2}$$

Since the A-body wave function is a sum of Faddeev components of all pairs, it
includes all pair-wise correlations. In terms of the Faddeev components, the A-body

Schrödinger equation can be written as

$$(T - E_T)\phi_{ij}(\vec{R}, \vec{r}_{ij}, \xi) = -V(\vec{r}_{ij}) \sum_{p,q>p}^{A} \phi_{pq}(\vec{R}, \vec{r}_{pq}, \xi), \qquad (9.3)$$

where T is the total kinetic energy operator and E_T is the total energy of the system. Equation (9.3) is equivalent to the A-body Schrödinger equation, as can be seen by summing both sides over all (ij) pairs and using Eq. (9.1). Corresponding equation for the relative motion is [see Eq. (8.16)]

$$(T' - E)\Phi_{ij}(\vec{r}_{ij}, \xi) = -V(\vec{r}_{ij})\Psi \qquad (9.4)$$

where Φ_{ij} and Ψ are the corresponding relative wave functions (after factoring out the wave function for the CM motion, which is a function of $\vec{R}$ only), E is the relative energy, and the relative kinetic energy operator is, using Eqs. (4.8) and (4.9)

$$T' = -\frac{\hbar^2}{m} \sum_{i=1}^{N} \nabla_{\xi_i}^2 \qquad (N = A - 1)$$

$$= -\frac{\hbar^2}{m} \left[\frac{\partial^2}{\partial \xi^2} + \frac{3A - 4}{\xi} \frac{\partial}{\partial \xi} - \frac{\mathcal{L}_{3N}^2(\Omega_{3N})}{\xi^2} \right]. \qquad (9.5)$$

We assume that the total orbital angular momentum $\vec{l}$ of the system is a good quantum number. For the (ij) Faddeev component, we can choose $\vec{\xi}_N = \vec{r}_{ij}$. Then Φ_{ij} is a function of $\vec{\xi}_N$ and ξ only and is independent of $\{\vec{\xi}_1, \ldots, \vec{\xi}_{N-1}\}$. Then all quantum numbers associated with these variables are zero [see Eq. (7.4)] for Φ_{ij}. Thus $\vec{l} = \vec{l}_N$ and $V(\vec{r}_{ij}) = V(r_{ij})$. Hence, we can expand Φ_{ij} in the PH subset $\{\mathcal{P}_{2K+l}^{lm}(\Omega_{3N}^{(ij)})\}$

$$\Phi_{ij}(\vec{r}_{ij}, \xi) = \xi^{-(3N-1)/2} F_{ij}(\vec{r}_{ij}, \xi) = \xi^{-(3N-1)/2} \sum_K \mathcal{P}_{2K+l}^{lm}(\Omega_{3N}^{(ij)}) u_K^l(\xi). \qquad (9.6)$$

The factor $\xi^{-(3N-1)/2}$ is introduced to remove the first derivative with respect to ξ in Eq. (9.4). We introduce a hyperangle ϕ through $r_{ij} = \xi \cos \phi$. Then the variables for the (ij)-partition become

$$(\xi, \Omega_{3N}) = (\xi, \phi, \omega_{ij}, \Omega_{3(N-1)}), \qquad (9.7)$$

where ω_{ij} represents the spherical polar angles of $\vec{r}_{ij}$. Finally, introducing a variable $z = \cos 2\phi$ (such that $r_{ij} = \xi \sqrt{\frac{1+z}{2}}$), a recurrence relation for the grand orbital operator is given by [using Eq. (4.9)]

$$\mathcal{L}_{3N}^2(\Omega_{3N}) = 4(1 - z^2)\frac{\partial^2}{\partial z^2} + 6[2 - N(1 + z)]\frac{\partial}{\partial z}$$
$$+ 2\frac{\hat{l}^2(\omega_{ij})}{1 + z} + 2\frac{\mathcal{L}_{3(N-1)}^2(\Omega_{3(N-1)})}{1 - z}. \tag{9.8}$$

In Chap. 7, we saw that the PH is an eigen function of $\mathcal{L}_{3N}^2(\Omega_{3N})$, corresponding to zero eigenvalue of $\mathcal{L}_{3(N-1)}^2(\Omega_{3(N-1)})$

$$\mathcal{L}_{3(N-1)}^2(\Omega_{3(N-1)})\mathcal{P}_{2K+l}^{lm}(\Omega_{3N}^{(ij)}) = 0, \tag{9.9}$$

such that $\mathcal{P}_{2K+l}^{lm}(\Omega_{3N}^{(ij)})$ is an eigenfunction of $\hat{l}^2(\omega_{ij})$ corresponding to eigenvalue $l(l + 1)$. Using Eqs. (7.5), (7.6), (4.15), and (4.16), the expression for this PH is

$$\mathcal{P}_{2K+l}^{lm}(\Omega_{3N}^{(ij)}) = N_K^l \, Y_{l,m}(\omega_{ij}) \, (\cos\phi)^l \, P_K^{\alpha,\beta}(z), \tag{9.10}$$

where $\alpha = (3A - 8)/2$ and $\beta = l + \frac{1}{2}$ and $P_K^{\alpha,\beta}(z)$ is the Jacobi polynomial. The constant N_K^l can be determined from the normalization condition

$$\int \mathcal{P}_{2K+l}^{lm}{}^*(\Omega_{3N}^{(ij)})\mathcal{P}_{2K'+l'}^{l'm'}(\Omega_{3N}^{(ij)})d\Omega_{3N}^{(ij)} = \delta_{KK'}\delta_{ll'}\delta_{mm'}. \tag{9.11}$$

Substituting Eq. (9.6) in Eq. (9.4) and using Eqs. (9.5), (9.8), and (9.9), we get

$$\left(\frac{\hbar^2}{m}\nabla_l^2 + E\right)F_{ij}(\vec{r}_{ij}, \xi) = V\left(\xi\sqrt{\frac{1 + z}{2}}\right)\sum_{p,q>p}F_{pq}(\vec{r}_{pq}, \xi), \tag{9.12}$$

where

$$\nabla_l^2 = \frac{\partial^2}{\partial\xi^2} - \frac{L(L + 1)}{\xi^2} + \frac{4}{\xi^2}\frac{1}{W_l(z)}\frac{\partial}{\partial z}(1 - z^2)W_l(z)\frac{\partial}{\partial z}, \tag{9.13}$$

with $L = l + (3A - 6)/2 = \alpha + l + 1$ and $W_l(z)$ being the weight function of Jacobi polynomial given by

$$W_l(z) = (1 - z)^\alpha(1 + z)^\beta. \tag{9.14}$$

From Eqs. (9.6) and (9.10), we can write

$$F_{ij}(\vec{r}_{ij}, \xi) = \sum_K \mathcal{P}_{2K+l}^{lm}(\Omega_{3N}^{(ij)})u_K^l(\xi) = Y_{l,m}(\omega_{ij})P_l(z_{ij}, \xi), \tag{9.15}$$

where z_{ij} now refers to the z-value for the ij-partition (the partition in which the ij-pair interacts) and $P_l(z, \xi)$ is given by

$$P_l(z, \xi) = \sum_K N_K^l \left(\frac{1+z}{2}\right)^{l/2} P_K^{\alpha,\beta}(z) \, u_K^l(\xi). \tag{9.16}$$

Similarly, we have for the pq-partition

$$F_{pq}(\vec{r}_{pq}, \xi) = \sum_K \mathcal{P}_{2K+l}^{lm}(\Omega_{3N}^{(pq)}) u_K^l(\xi) = Y_{l,m}(\omega_{pq}) P_l(z_{pq}, \xi), \tag{9.17}$$

If we project Eq. (9.12) on a particular $\mathcal{P}_{2K+l}^{lm}(\Omega_{3N}^{(ij)})$, then matrix elements of the form

$$\langle \mathcal{P}_{2K+l}^{lm}(\Omega_{3N}^{(ij)}) | V(\vec{r}_{ij}) | \sum_{p,q>p} F_{pq}(\vec{r}_{pq}, \xi) \rangle$$

will appear. Now $\mathcal{P}_{2K+l}^{lm}(\Omega_{3N}^{(ij)}) V(\vec{r}_{ij})$ is a function of $\vec{r}_{ij}$ and ξ only. Hence, it can be fully expanded in the basis $\{\mathcal{P}_{2K+l}^{lm}(\Omega_{3N}^{(ij)})\}$. However, $\sum_{p,q>p} F_{pq}(\vec{r}_{pq}, \xi)$ is a function of *all pair separations* $\vec{r}_{pq}$, including $\vec{r}_{ij}$. Hence, it cannot be fully expanded in the basis $\{\mathcal{P}_{2K+l}^{lm}(\Omega_{3N}^{(ij)})\}$. However, for the above matrix element, the projection of $\sum_{p,q>p} F_{pq}(\vec{r}_{pq}, \xi)$ on the $\vec{r}_{ij}$ space will only contribute, the residual part being orthogonal to $\mathcal{P}_{2K+l}^{lm}(\Omega_{3N}^{(ij)})$. This is similar to the fact that the dot product of a vector $\vec{A}$, which lies entirely in the (xy) plane with a three-dimensional vector $\vec{B}$ can be replaced by the dot product of $\vec{A}$ with the projection of $\vec{B}$ on the (xy) plane.

Hence, in Eq. (9.12), we replace $\sum_{p,q>p} F_{pq}(\vec{r}_{pq}, \xi)$ by its projection on the $\vec{r}_{ij}$-space

$$\left(\frac{\hbar^2}{m}\nabla_l^2 + E\right) F_{ij}(\vec{r}_{ij}, \xi) = V\left(\xi\sqrt{\frac{1+z}{2}}\right) \hat{\mathbf{P}}_{\vec{r}_{ij}} \sum_{p,q>p} F_{pq}(\vec{r}_{pq}, \xi), \tag{9.18}$$

where $\hat{\mathbf{P}}_{\vec{r}_{ij}}$ is the operator for projection on to the $\vec{r}_{ij}$-space. The projection of $F_{pq}(\vec{r}_{pq}, \xi)$ on the $\vec{r}_{ij}$ space, viz. $\hat{\mathbf{P}}_{\vec{r}_{ij}} F_{pq}(\vec{r}_{pq}, \xi)$, is done first by expanding $F_{pq}(\vec{r}_{pq}, \xi)$ in the corresponding PH basis $\{\mathcal{P}_{2K+l}^{l,m}(\Omega_{3N}^{pq})\}$ as in Eq. (9.17) and then applying $\hat{\mathbf{P}}_{\vec{r}_{ij}}$ which can be written as

$$\hat{\mathbf{P}}_{\vec{r}_{ij}} = \hat{\mathbf{P}}_{\Omega_{3N}^{(ij)}} \hat{\mathbf{P}}_{z_{ij}}, \tag{9.19}$$

where $\hat{\mathbf{P}}_{\Omega_{3N}^{(ij)}}$ projects a PH for the pq-partition on to a PH for the ij-partition, viz.

$$\hat{\mathbf{P}}_{\Omega_{3N}^{(ij)}} = \left| \mathcal{P}_{2K+l}^{lm}(\Omega_{3N}^{(ij)}) \right\rangle \left\langle \mathcal{P}_{2K+l}^{lm}(\Omega_{3N}^{(ij)}) \right|. \tag{9.20}$$

The bra vector on the right acts on a PH for the pq-partition, giving an inner product, which is the projection amplitude of the PH for the ij-partition, the latter being represented by the ket vector. Note that the hyperangular momentum quantum number K is conserved for different partitions. The operator $\hat{\mathbf{P}}_{z_{ij}}$ projects a function of z_{pq} on to the z_{ij}-space. In the expression for a PH, Eq. (9.10), z appears as the argument of a Jacobi polynomial, which satisfies the orthonormality relation

$$\int_{-1}^{1} [P_K^{\alpha,\beta}(z)]^* \, P_{K'}^{\alpha,\beta}(z) \, W_l(z) dz = h_K^{\alpha,\beta} \, \delta_{KK'}, \tag{9.21}$$

where $h_K^{\alpha,\beta}$ is the norm of Jacobi polynomials, given by Eq. (3.20). The closure relation satisfied by the Jacobi polynomials is

$$\sum_K \frac{1}{h_K^{\alpha,\beta}} P_K^{\alpha,\beta}(z) \, [P_K^{\alpha,\beta}(z')]^* \, W_l(z') = \delta(z - z'). \tag{9.22}$$

Hence, the projection operator $\hat{\mathbf{P}}_z$, which acting on a function of z' projects it on to the z-space, is given by

$$\hat{\mathbf{P}}_z = \sum_K P_K^{\alpha,\beta}(z) \frac{1}{h_K^{\alpha,\beta}} \int_{-1}^{1} dz' [P_K^{\alpha,\beta}(z')]^* W_l(z'). \tag{9.23}$$

With z and z' replaced by z_{ij} and z_{pq}, respectively, we have

$$\hat{\mathbf{P}}_{z_{ij}} = \sum_K P_K^{\alpha,\beta}(z_{ij}) \frac{1}{h_K^{\alpha,\beta}} \int_{-1}^{1} dz_{pq} [P_K^{\alpha,\beta}(z_{pq})]^* W_l(z_{pq}). \tag{9.24}$$

This operator acting on a function of z_{pq} mathematically gives a function of z_{ij}, with no change in the reference partition (which is pq-partition) for the PH, while $\hat{\mathbf{P}}_{\Omega_{3N}^{(ij)}}$ projects it further on to a different reference partition (ij-partition in this case) for the PH. The scalar variable z_{pq} in Eq. (9.24) is just a variable of integration of a definite integral, and hence does not refer to any particular partition. As the Jacobi polynomials are real, the complex conjugation may be dropped.

To evaluate $\hat{\mathbf{P}}_{\vec{r}_{ij}} \sum_{p,q>p} F_{pq}(\vec{r}_{pq}, \xi)$ of Eq. (9.18), we first apply $\hat{\mathbf{P}}_{z_{ij}}$ on $\sum_{p,q>p} F_{pq}$ $(\vec{r}_{pq}, \xi)$, using Eqs. (9.17) and (9.24)

$$\hat{\mathbf{P}}_{z_{ij}} \sum_{p,q>p} F_{pq}(\vec{r}_{pq}, \xi)$$

$$= \sum_{p,q>p} Y_{lm}(\omega_{pq}) \, \hat{\mathbf{P}}_{z_{ij}} \, P_l(z_{pq}, \xi)$$

$$= \sum_{p,q>p} Y_{lm}(\omega_{pq}) \sum_K P_K^{\alpha,\beta}(z_{ij}) \frac{1}{h_K^{\alpha,\beta}} \int_{-1}^1 dz_{pq} [P_K^{\alpha,\beta}(z_{pq})]^* W_l(z_{pq}) P_l(z_{pq}, \xi). \quad (9.25)$$

Next we apply the operator $\hat{\mathbf{P}}_{\Omega_{3N}^{(ij)}}$ from Eq. (9.20) on Eq. (9.25). The result of this projection is given in terms of $\mathcal{P}_{2K+l}^{lm}(\Omega_{3N}^{(ij)})$, which contains $Y_{lm}(\omega_{ij})$. As mentioned earlier, z_{pq} in Eq. (9.25) is a variable of integration of a definite integral and is not affected by the $\hat{\mathbf{P}}_{\Omega_{3N}^{(ij)}}$ projection (we will later replace this variable simply by z'). Thus we get

$$\hat{\mathbf{P}}_{\Omega_{3N}^{(ij)}} \hat{\mathbf{P}}_{z_{ij}} \sum_{p,q>p} F_{pq}(\vec{r}_{pq}, \xi)$$

$$= \sum_K \sum_{p,q>p} \left\langle \mathcal{P}_{2K+l}^{lm}(\Omega_{3N}^{(ij)}) \middle| \mathcal{P}_{2K+l}^{lm}(\Omega_{3N}^{(pq)}) \right\rangle$$

$$\times Y_{lm}(\omega_{ij}) P_K^{\alpha,\beta}(z_{ij}) \frac{1}{h_K^{\alpha,\beta}} \int_{-1}^1 dz_{pq} [P_K^{\alpha,\beta}(z_{pq})]^* W_l(z_{pq}) P_l(z_{pq}, \xi)$$

$$= Y_{lm}(\omega_{ij}) \sum_K f_{Kl}^2 P_K^{\alpha,\beta}(z_{ij}) \frac{1}{h_K^{\alpha,\beta}} \int_{-1}^1 dz_{pq} [P_K^{\alpha,\beta}(z_{pq})]^* W_l(z_{pq}) P_l(z_{pq}, \xi) \quad (9.26)$$

In the above, we have used Eq. (7.22): $f_{Kl}^2 = \sum_{p,q>p} \left\langle \mathcal{P}_{2K+l}^{lm}(\Omega_{3N}^{(ij)}) \middle| \mathcal{P}_{2K+l}^{lm}(\Omega_{3N}^{(pq)}) \right\rangle$, which is a constant given by Eq. (7.23). Replacing the variable of integration z_{pq} by z' in Eq. (9.26), using Eq. (9.19) and noting that Jacobi polynomials are real, we have

$$\hat{\mathbf{P}}_{\vec{r}_{ij}} \sum_{p,q>p} F_{pq}(\vec{r}_{pq}, \xi)$$

$$= Y_{lm}(\omega_{ij}) \sum_K f_{Kl}^2 P_K^{\alpha,\beta}(z_{ij}) \frac{1}{h_K^{\alpha,\beta}} \int_{-1}^1 dz' P_K^{\alpha,\beta}(z') W_l(z') P_l(z', \xi). \quad (9.27)$$

In order to separate the term $F_{ij}(\vec{r}_{ij}, \xi)$ from the sum over all partitions, we add and subtract 1 from f_{Kl}^2 on the right side of Eq. (9.27) and use the closure relation (9.22) of Jacobi polynomials to get

$$\hat{\mathbf{P}}_{\vec{r}_{ij}} \sum_{p,q>p} F_{pq}(\vec{r}_{pq}, \xi) = Y_{lm}(\omega_{ij}) \Bigg[P_l(z_{ij}, \xi)$$

$$+ \sum_K (f_{Kl}^2 - 1) P_K^{\alpha,\beta}(z_{ij}) \frac{1}{h_K^{\alpha,\beta}} \int_{-1}^{1} dz' P_K^{\alpha,\beta}(z') W_l(z') P_l(z', \xi) \Bigg].$$

$$= Y_{lm}(\omega_{ij}) \Bigg[P_l(z_{ij}, \xi) + \int_{-1}^{1} dz' f_l(z_{ij}, z') P_l(z', \xi) \Bigg], \qquad (9.28)$$

where we define

$$f_l(z, z') = \sum_K (f_{Kl}^2 - 1) P_K^{\alpha,\beta}(z) \frac{1}{h_K^{\alpha,\beta}} P_K^{\alpha,\beta}(z') W_l(z'). \qquad (9.29)$$

Using Eq. (9.15) in Eq. (9.18), substituting Eq. (9.28), eliminating $Y_{lm}(\omega_{ij})$, and replacing z_{ij} by z, we have

$$\left(\frac{\hbar^2}{m} \nabla_l^2 + E \right) P_l(z, \xi) = V \left(\xi \sqrt{\frac{1+z}{2}} \right) \Pi_l(z, \xi)$$

$$= V \left(\xi \sqrt{\frac{1+z}{2}} \right) \Bigg[P_l(z, \xi) + \int_{z'=-1}^{1} f_l(z, z') P_l(z', \xi) dz' \Bigg],$$

$$(9.30)$$

where ∇_l^2 is given by Eq. (9.13) and the kernel function $f_l(z, z')$ is given by Eq. (9.29). A different but equivalent treatment has been adopted in Ref. [3], which results in a closed analytic expression for the kernel function for $l = 0$.

Since the IDE is derived from the PHEM, they are equivalent, provided all K-partial waves are included in the latter. The advantage of using the IDE is that it does not involve K-partial waves, as the kernel function can be obtained directly; hence, one does not have to repeatedly solve a system of gradually increasing number of CDE and look for convergence. The infinite sum over K for the kernel function in Eq. (9.29) appears due to the fact that we derived the IDE from the PHEM. If we try to evaluate the kernel function using a truncated sum over K, we will again have to solve the IDE repeatedly with an increasing upper limit in K value in Eq. (9.29). Indeed, this is worthwhile in few-nucleon nuclei [1]. However, in some cases, the kernel function may be obtained directly in closed form, without any reference to K-partial waves [3]. In specific cases, it may be possible to evaluate the infinite sum over K in Eq. (9.29) resulting in a closed form [10]. With a known $f_l(z, z')$, one has to solve the IDE only once. Thus in contrast with the PHEM, the IDE can be solved only once, without the need to solve successively bigger sets of equations to check for convergence. Usefulness of the IDE depends on evaluation of an accurate kernel function in a closed form. One great advantage of the IDE approach is that, unlike the HHEM or Faddeev–Yakubovsky equation methods, its complexity does not increase with A.

9.2 Applications of IDE

The IDE, like the HHEM and PHEM, was originally applied to nuclear systems, with relatively small number of nucleons interacting through s-projected potentials by Fabre and his collaborators. Later, Adam and Sofianos used a mathematical trick in the large A limit to evaluate the K sum in the kernel function in a closed form and then applied the technique to Bose–Einstein condensates (BEC) with a fairly large number of bosons. In the following we will discuss these.

9.2.1 Nuclear Systems

The simplest nuclear application is the one for the trinucleon systems, $A = 3$, corresponding to $\alpha = \frac{1}{2}$. In Chap. 5, we saw that the totally antisymmetric ground state has a large contribution from the space totally symmetric S-state ($l = 0$, hence $\beta = \frac{1}{2}$) of the system, whose isospin–spin part is totally antisymmetric. The kernel function for this state can be obtained in a closed form [3]. The IDE for this state has the form [1, 3]

$$\left(\frac{\hbar^2}{m}\nabla_0^2 + E\right)P_0(z, \xi) = V\left(\xi\sqrt{\frac{1+z}{2}}\right)\Pi_0(z, \xi)$$

$$= V\left(\xi\sqrt{\frac{1+z}{2}}\right)\left[P_0(z, \xi) + \frac{2}{\sqrt{3(1-z^2)}}\int_{z_-}^{z_+} P_0(z', \xi)dz'\right],$$

$$(9.31)$$

where $z_\pm = \frac{1}{2}(-z \pm \sqrt{3(1-z^2)})$. The Faddeev component is given by

$$\Phi_{ij}(\vec{\xi}_1, \ldots, \vec{\xi}_N) = \xi^{-\frac{5}{2}}\sum_K \mathcal{P}_{2K}^{00}(\Omega_6^{(ij)})u_K^0(\xi) = \xi^{-\frac{5}{2}}P_0(z, \xi),$$

$$(9.32)$$

with $z = 2\frac{r_{ij}^2}{\xi^2} - 1$. It can be shown [1] that Eq. (9.31) has the same form as the three nucleon Faddeev equation for the S state. To test the validity of the IDE, Fabre, Fiedeldey, and Sofianos [3] solved the IDE for three- and four-nucleon nuclei using adiabatic approximation (see Chap. 10, Sect. 10.2.3 for a description of adiabatic approximation) for the S-state with a number of standard central nucleon–

nucleon potentials and compared the results with other accurate calculations, including Faddeev–Yakubovsky equation method. Binding energy of ^{3}H calculated by the IDE approach (interpolated between extreme and uncoupled adiabatic approximations) agrees within 0.1 % with most other accurate calculations. In Table 9.1, we present selected results for comparison of ^{3}H binding energy by different methods (ETBM stands for the variational method with correlation functions). Table 9.2 presents binding energies of ^{4}He nucleus (without Coulomb interaction) calculated by interpolated IDE for selected potentials and comparison with other calculations (ETBM and GFMC stand for variational calculation with correlation function and Green's function Monte Carlo calculations, respectively). We can see from these tables that although the IDE results are quite reliable for the trinucleon, they are less reliable for the four-nucleon system (differing by up to 0.7 MeV). It is not surprising that the IDE compares worse with other accurate methods for the four-nucleon system, since higher than two-body correlations contribute non-negligibly for dense four-body systems.

For the trinucleon system interacting through central (including spin-orbit term) and tensor interactions, three symmetry components, viz., S, S′, and D states,

Table 9.1 Comparison of trinucleon binding energy (in MeV) by different methods using different potentials [3]

Potential	Interpolated IDE	HH	ETBM	Faddeev
Volkov	8.47	8.465	8.460	
EH (S4)	7.08	7.05	7.04	
AT (S3)	6.69	6.695	6.677	6.696
MTV Erens	7.78	7.783	7.778	
MTV Friar	7.73			7.736
MTV Zabol	8.25			8.253

Table 9.2 Comparison of ^{4}He binding energy (in MeV, without Coulomb interaction) by different methods using different potentials [3]

Potential	Interpolated IDE	HH	Other results	Method
Volkov	30.40	30.40	30.32	ETBM
GPDT	18.32	18.29		
EH (S4)	28.71	27.9	28.18	ETBM
AT (S3)	26.92	26.0	26.47	ETBM
MTV Erens	29.42			
MTV Friar	29.34			
MTV Zabol	30.63		31.36	Faddeev
			31.3±0.2	GFMC

contribute (see Chap. 5). In this case, there will be three coupled integro-differential equations satisfied by the three components P^S, $P^{S'}$, and P^D of Φ_{ij} [1]

$$\left(\frac{\hbar^2}{m}\nabla_0^2 + E\right)P_0^S(z, \xi) = \frac{1}{2}(V^{1+} + V^{3+})\Pi_0^S(z, \xi) + \frac{1}{2}(V^{1+} - V^{3+})\Pi_0^{S'}(z, \xi)$$
$$- 2V_T^+(1 + z)\Pi_2^D(z, \xi)$$

$$\left(\frac{\hbar^2}{m}\nabla_0^2 + E\right)P_0^{S'}(z, \xi) = \frac{1}{2}(V^{1+} + V^{3+})\Pi_0^{S'}(z, \xi) + \frac{1}{2}(V^{1+} - V^{3+})\Pi_0^S(z, \xi)$$
$$+ 2V_T^+(1 + z)\Pi_2^D(z, \xi)$$

$$\left(\frac{\hbar^2}{m}\nabla_2^2 + E\right)P_2^D(z, \xi) = (V^{3+} - 2V_T^+)\Pi_2^D(z, \xi)$$
$$+ \frac{2}{1 + z}V_T^+\left(\Pi_0^{S'}(z, \xi) - \Pi_0^S(z, \xi)\right), \tag{9.33}$$

where V^{1+}, V^{3+}, and V_T^+ are singlet even, triplet even, and tensor even potentials, respectively, and the argument of each potential is $\xi\sqrt{(1 + z)/2}$. Corresponding Π-functions are given by

$$\Pi_0^S(z, \xi) = P_0^S(z, \xi) + \frac{2}{\sqrt{3(1 - z^2)}}\int_{z_-}^{z_+} P_0^S(z', \xi)dz'$$

$$\Pi_0^{S'}(z, \xi) = P_0^{S'}(z, \xi) - \frac{1}{\sqrt{3(1 - z^2)}}\int_{z_-}^{z_+} P_0^{S'}(z', \xi)dz'$$

$$\Pi_2^D(z, \xi) = P_2^D(z, \xi) - \int_{-1}^{+1} f_D(z, z')P_2^D(z', \xi)dz', \tag{9.34}$$

where $f_D(z, z')$ is given by Eq. (9.29) and $z_{\pm} = \frac{1}{2}(-z \pm \sqrt{3(1 - z^2)})$. The IDE for three- and four-nucleon systems including S and S' states for central forces have been solved in the adiabatic approximation by Oehm et al. [5]. The same systems including S, S', and D states were also treated by IDE for realistic nucleon–nucleon potentials like Reid soft-core (RSC) potential and compared with Faddeev–Yakubovsky calculations [6, 7]. Good agreement was obtained, establishing the dominance of two-body correlations over higher-body correlations in these systems, even for realistic forces. In actual calculations, the hypercentral average has been added to and subtracted from the central two-body potentials appearing in the IDE [Eqs. (9.31) and (9.33)], to accelerate the rate of convergence. We skip the details here, for separate discussion in Sect. 10.2 of Chap. 10.

The IDE has also been applied by Fabre, Sofianos, and Adam to larger nuclear systems: closed shell nuclei ^{4}He, ^{16}O and ^{40}Ca as also open-shell nucleus ^{10}B [8]. In these calculations, Coulomb repulsion has been included. More recently, Sofianos, Adam, and Belyaev used the IDE for an α particle description of $A\alpha$ nuclei [9]. These successful applications demonstrate the usefulness of the IDE for nuclear systems containing up to 40 nucleons. As we have already discussed, the advantage of the

IDE is that its complexity does not increase with the number of nucleons, unlike other few-body methods. However, nuclear systems are dense and the assumption that only two-body correlations dominate will not be valid for larger systems. By contrast, the Bose–Einstein condensate is very dilute and such an assumption is valid for much larger number of bosons, as we discuss in the next subsection.

9.2.2 Bose–Einstein Condensates

Bose–Einstein condensates (BEC) are achieved in the laboratory by trapping a large number of bosons (usually neutral odd-mass alkali atoms) by a harmonic oscillator potential and cooling down to a few hundred nano-Kelvin. The BEC is deliberately kept extremely dilute to eliminate three- and higher-body collisions, which lead to molecule formation and consequent depletion (see Chap. 8). The diluteness assures that only two-body forces are active and the wave function contains no more than two-body correlations. The conditions are ideal for decomposing the many-body wave function in Faddeev components and expanding each Faddeev component in the potential harmonics basis. Hence, it is no wonder that the PHEM finds a successful application in the BEC (Chap. 8). For the same reason, the IDE is expected to be equally good for the BEC. In Sect. 9.1, we saw that the kernel function of IDE involves the Jacobi polynomial $P_K^{\alpha,\beta}(z)$ and its weight function $W_l(z) = (1 - z)^\alpha (1 + z)^\beta$ with $\alpha = (3A - 8)/2$, which increases rapidly with A. For large $A \sim 100$, the weight function becomes extremely sharply peaked at $z = \frac{\beta - \alpha}{\alpha + \beta}$ (which is slightly greater than -1), has a maximum $\sim 2^\alpha$, and then decreases very rapidly within a tiny interval, as z increases. The sharpness of the peak and its peak-value increase very fast with increase in A. This causes a serious numerical problem. Adam and Sofianos [10] used a mathematical limit as $A \to \infty$ to convert Jacobi polynomial and its weight function to associated Laguerre polynomial and its weight function. The latter are very smooth functions and the kernel of the resulting IDE has a simple analytic expression. This has been used for A up to 100 and the results agree well with other accurate calculations. In the following, we follow Ref. [10] to obtain the modified IDE and its kernel function. However, in order to keep the treatment in the same line as in Sect. 9.1, we leave out a hypercentral average term, which was added to and subtracted from the central two-body potential to make the convergence faster. The latter will be discussed separately in Chap. 10, Sect. 10.2.

The IDE and its kernel function for the general case are given by Eqs. (9.30) and (9.29), respectively. For the ground state of the BEC, $l = 0$ ($\beta = \frac{1}{2}$) and there is a confining potential $V_{trap} = \sum_{i=1}^{A} \frac{1}{2}m\omega^2 r_i^2 = \frac{1}{4}m\omega^2\xi^2$. Hence,

$$\left(\frac{\hbar^2}{m}\nabla_0^2 - V_{trap}(\xi) + E\right)P_0(z, \xi) = V\left(\xi\sqrt{\frac{1+z}{2}}\right)$$
$$\times \left[P_0(z, \xi) + \int_{-1}^{1} f_0(z, z')P_0(z', \xi)dz'\right].$$
(9.35)

As discussed above, $P_K^{\alpha,\beta}(z)$ and $W_l(z)$ appearing in Eq. (9.29) become very difficult to handle numerically for large A. We introduce a new variable ζ in place of z through

$$r_{ij} = \frac{\zeta\xi}{\sqrt{\alpha}},$$
(9.36)

such that $z = 2\zeta^2/\alpha - 1$. Then use the mathematical limit [11]

$$\lim_{\alpha\to\infty} P_K^{\alpha,\frac{1}{2}}\left(2\frac{r_{ij}^2}{\xi^2} - 1\right) = (-1)^K L_K^{\frac{1}{2}}\left(\frac{\alpha r_{ij}^2}{\xi^2}\right),$$
$$= (-1)^K L_K^{\frac{1}{2}}(\zeta^2)$$
(9.37)

where $L_K^{\frac{1}{2}}(\zeta^2)$ is the associated Laguerre function. In the same limit, the weight function of the Jacobi polynomial becomes (for $l = 0$)

$$\lim_{\alpha\to\infty} W_l(z) = C_W \frac{2^{\alpha+\frac{1}{2}}}{\alpha^{\frac{1}{2}}}\zeta e^{-\zeta^2},$$
(9.38)

where C_W is the normalization for the weight function. Note that in this limit, the weight function has the form appropriate for $L_K^{\frac{1}{2}}(\zeta^2)$. Substituting

$$P_0(\zeta, \xi) = \frac{e^{\zeta^2/2}}{\zeta}Q(\zeta, \xi)$$
(9.39)

in Eq. (9.35), we have

$$\left[H_\xi + \frac{4}{\xi^2}H_\zeta - E\right]Q(\zeta, \xi) = -V\left(\frac{\zeta\xi}{\sqrt{\alpha}}\right)\left[Q(\zeta, \xi) + \int_0^{\sqrt{\alpha}} \mathcal{F}_0(\zeta, \zeta')Q(\zeta', \xi)d\zeta'\right],$$
(9.40)

where

$$H_\xi = \frac{\hbar^2}{m} \left[-\frac{\partial^2}{\partial \xi^2} + \frac{\mathcal{L}(\mathcal{L}+1)}{\xi^2} \right] + V_{\text{trap}}(\xi) \tag{9.41}$$

and

$$H_\zeta = \frac{\hbar^2}{m} \frac{\alpha}{4} \left[-\frac{\partial^2}{\partial \zeta^2} + \zeta^2 - 3 \right]. \tag{9.42}$$

In the limit $\alpha \to \infty$, the kernel function can be put in a closed form [10]

$$\mathcal{F}_0(\zeta, \zeta') = \frac{2(A-2)}{\sqrt{3}} \left[\left(A - 3 - \frac{2}{3} \left(\zeta^2 - \frac{3}{2} \right) \left(\zeta'^2 - \frac{3}{2} \right) \right) \zeta \zeta' e^{-(\zeta^2 + \zeta'^2)/2} \right.$$
$$\left. + \frac{4}{\sqrt{3}} \left[e^{-[5(\zeta-\zeta')+2\zeta\zeta']/6} - e^{-[5(\zeta+\zeta')-2\zeta\zeta']/6} \right] \right]. \tag{9.43}$$

Adam and Sofianos [10] solved Eqs. (9.40)–(9.43) for A atoms of ^{87}Rb confined in the harmonic oscillator potential having circular frequency $\omega = 2\pi\nu$ with $\nu = 200\,\text{Hz}$, simulating the JILA trap. Energy and length were expressed in units of $\hbar\omega$ and $\sqrt{\frac{\hbar}{m\omega}}$ and referred to as oscillator units (o.u.) of energy and length, respectively. Two-body interaction was represented by two simple semi-realistic potentials:

1. Potential V1: Gaussian potential

$$V(r_{ij}) = V_0 \exp\left(-\frac{r_{ij}^2}{r_0^2}\right)$$

   with $V_0 = 3.1985 \times 10^6$ o.u. and $r_0 = 0.005$ o.u.
2. Potential V2: Sech-squared potential:

$$V(r_{ij}) = V_0 \operatorname{sech}^2\left(\frac{r_{ij}}{r_0}\right)$$

   with $V_0 = 1.81847 \times 10^7$ o.u. and $r_0 = 0.001$ o.u.

Table 9.3 Comparison of ground state energies (in o.u.) of a BEC containing A atoms of ^{87}Rb, calculated by IDE, PHEM, and DMC

A	Potential V1		Potential V2		
	IDE	PHEM	IDE	PHEM	DMC
3	6.009	4.500			
5	7.758	7.505			
10	15.003	15.034	15.143	15.1490	15.1539
20	30.001	30.107	30.625	30.6209	30.639
35	52.501	52.768			
50			78.701	78.8704	
100			165.038	164.907	

The parameters of the potentials chosen correspond to $a_s = 100$ Bohr. They solved the IDE using adiabatic approximation and also adding and subtracting the hypercentral average (see Chap. 10, Sects. 10.2.3 and 10.2.2, respectively). The results are quoted in Table 9.3.

One can notice from this table that the ground state energies calculated by the IDE are quite close to those by the PHEM and are also close to the DMC results, where the latter is available. The small differences arise mostly from numerical errors. Thus the IDE is an important alternative method, which is both simple and at the same time, its complexity does not increase with the number of particles. The IDE has one advantage over the PHEM: the wave function calculated by the IDE is better in the asymptotic region, as we discussed earlier. Although the IDE method is less popular than the HHEM, it has been applied to atomic three-body systems by Sultanov and Guster [12].

References

1. Fabre de la Ripelle, M.: Few-Body Syst. **1**, 181 (1986)
2. Fabre de la Ripelle, M., Fiedeldey, H., Sofianos, S.A.: Few-Body Syst. Suppl. **2**, 493 (1987)
3. Fabre de la Ripelle, M., Fiedeldey, H., Sofianos, S.A.: Phys. Rev. C **38**, 449 (1988)
4. Fabre de la Ripelle, M., Fiedeldey, H., Sofianos, S.A.: Few-Body Syst. **6**, 157 (1989)
5. Oehm, W., Sofianos, S.A., Fiedeldey, H., Fabre de la Ripelle, M.: Phys. Rev. C **43**, 25 (1991)
6. Oehm, W., Sofianos, S.A., Fiedeldey, H., Fabre de la Ripelle, M.: Phys. Rev. C **44**, 81 (1991)
7. Fabre de la Ripelle, M., Braun, M., Sofianos, S.A.: Prog. Theor. Phys. **98**, 1261 (1997)
8. Fabre de la Ripelle, M., Sofianos, S.A., Adam, R.M.: Ann. Phys. **316**, 107 (2005)
9. Sofianos, S.A., Adam, R.M., Belyaev, V.B.: Phys. Rev. C **84**, 064304 (2011)
10. Adam, R.M., Sofianos, S.A.: Phys. Rev. A **82**, 053635 (2010)
11. Abramowitz, M., Stegun, I.A.: Handbook of Mathematical Functions. Dover Publications Inc., New York (1972)
12. Sultanov, R.A., Guster, D.: J. Comput. Phys. **192**, 231 (2003)

Chapter 10
Computational Techniques

Abstract Some computational techniques are presented here. A method is presented for solution of a single differential eigenvalue equation, subject to boundary conditions at the origin and at infinity. Next solution of a system of coupled differential eigenvalue equations (CDEE) is discussed. First an exact numerical algorithm, viz., renormalized Numerov (RN) method is presented. Next approximation methods are discussed. Introduction of a hypercentral average enhances the rate of convergence. Hyperspherical adiabatic approximation (HAA) reduces the CDEE to a single differential equation. Applicability, accuracy, and numerical procedure of HAA are discussed. Finally, a method is presented for handling tricky integrals (involving extremely fast changing integrand) in potential matrix element.

In actual applications of the hyperspherical harmonics techniques, one has to solve the equations and evaluate physical quantities numerically. Most of these calculations involve quite intricate numerical methods. Many of these calculations need special numerical techniques tailored to the particular need and are different from the standard ones. As an example, we see that the weight function of the Jacobi polynomial for a large number of particles varies extremely rapidly over an exceedingly small interval. Standard numerical quadratures involving this weight function in the integrand produces a very large error, needing a special treatment. As another example, one sees that the computation time required for evaluating a number of potential matrix elements accurately can be reduced drastically by solving a set of linear inhomogeneous equations, instead of using standard quadratures for the integrals.

In this chapter, we discuss the numerical techniques to handle such problems. For the sake of completeness, we also briefly discuss the numerical methods for solving more standard problems, like the eigen solution of a single differential equation in Sect. 10.1. Section 10.2 deals with solution of a set of coupled differential equations (CDE). Computation of potential matrix elements will be discussed in Sect. 10.3.

© Springer India 2016   141
T.K. Das, *Hyperspherical Harmonics Expansion Techniques*,
Theoretical and Mathematical Physics, DOI 10.1007/978-81-322-2361-0_10

10.1   Solution of a Single Differential Equation

A simple second order differential equation is the most common one. It results from
the Schrödinger equation for a single particle in a central field (Sect. 2.1) or a two-
body system with mutual central force (Sect. 2.2). Its general form is

$$\left[-\frac{\hbar^2}{2\mu} \frac{d^2}{dr^2} + \frac{\hbar^2}{2\mu} \frac{l(l+1)}{r^2} + V(r) - E \right] u(r) = 0, \qquad (10.1)$$

where μ is the mass of the single particle in the central field or the reduced mass of
the two-body system and l is the orbital angular momentum of the system. Systems
containing more than two particles (for which r is replaced by ξ in our notation), or
systems with noncentral interaction give rise to a set of CDE. An adiabatic approximation can reduce the CDE to a single differential equation of the above form. Thus
Eq. (10.1) as the ultimate form to be solved is quite common. Some text books of
quantum mechanics provide the basic discussion of the numerical procedure [1].
Here, we present a more detailed discussion.

For bound states in a potential $V(r)$ which vanishes at infinity, the energy eigenvalue E is negative. It is obtained by applying boundary conditions on the eigen
function $u(r)$ appropriate for a bound state, viz., $u(r)$ must vanish at $r = 0$ and
$r \to \infty$. For a numerical solution, $r = 0$ should be avoided, since the centrifugal
term diverges at this point. Hence, a small enough r value (say r_0) should be chosen
as the initial point. Also $r = \infty$ should be replaced by a large enough, but finite,
r value (say) r_f. Criterion for choosing r_f should be $|V(r_f)| < \epsilon$, where ϵ is the
error limit demanded for E. Next we should choose a matching point r_m, such that
$r_m > r_{min}$, where r_{min} is the position of the minimum of $V(r)$. To obtain the eigen
solution, one has to perform the following steps:

1. Divide the interval $[r_0, r_f]$ in two subintervals: $r_0 \le r \le r_m$ and $r_m \le r \le r_f$.
   Choose an initial trial energy $E = E_0 < 0$.
2. Integrate Eq. (10.1) point-by-point (with a selected r-step size h) in the forward
   direction of the first subinterval from $r = r_0$ to $r = r_m$, starting from the initial
   values [$u(0) = 0$ and $u'(0) \ne 0$ are replaced for finite r_0 by limiting solution of
   Eq. (10.1)]

$$u(r_0) \simeq C_L r_0^{(l+1)}, \qquad u'(r_0) = C_L(l+1)r_0^l,$$

   where C_L is an arbitrary constant. The standard procedure to solve the second
   order ordinary differential equation is to write it as a set of two coupled first-order
   differential equations and solve them by a standard algorithm like Runge–Kutta
   method [2, 3]. Store the ratio $D_L(E) = u'(r_m)/u(r_m)$.
3. Integrate Eq. (10.1) point-by-point in the second sub-interval backwards from
   $r = r_f$ to $r = r_m$, starting from the asymptotic values [obtained from asymptotic
   solution of Eq. (10.1)]

$$u(r_f) = C_R \exp(-\kappa r_f), \quad u'(r_f) = -\kappa C_R \exp(-\kappa r_f), \quad \kappa^2 = -\frac{2\mu E}{\hbar^2},$$

(where C_R is an arbitrary constant) by the same method. Store the ratio $D_R(E) = u'(r_m)/u(r_m)$.

4. Solve the equation of continuity of log-derivatives of wave function in the two subintervals at the match point

$$G(E) = D_L(E) - D_R(E) = 0, \tag{10.2}$$

by a suitable root finder algorithm like the bisection method or the Newton–Raphson method [2], to find E, which satisfies Eq. (10.2). As an example, we discuss here the Newton–Raphson method. For the chosen energy (E), $G(E)$ is in general not equal to zero. Calculate the correction to E as

$$\Delta E = -\frac{G(E)}{\frac{dG}{dE}}.$$

The local derivative $\frac{dG}{dE}$ can be approximately calculated from the stored $G(E)$ corresponding to the energy of the previous cycle. Then the trial energy is improved by replacing E by $E + \Delta E$ and steps (2), (3), and (4) are repeated until desired precision is achieved, i.e., $|\Delta E| < \epsilon$. The final value of E is the energy eigen value.

5. With this final E, Eq. (10.1) is once again solved from $r = r_0$ to $r = r_f$, subject to initial values in step (2), to obtain the wave function $u(r)$ in the entire chosen interval. It is then normalized to get the normalized eigen function.

The procedure outlined above is the basic algorithm for solving a single differential eigenvalue equation. The method can be refined in accordance with a particular situation. Some of these are discussed below.

- If $V(r)$ for $r \to 0$ has an r-dependence stronger than r^{-2}, then the appropriate $r \to 0$ solution of Eq. (10.1) is to be used in step (2) above. Similarly, if $V(r)$ does not vanish asymptotically (as in the case of a harmonic oscillator potential) we have to use appropriate asymptotic solution in step (3) above. Note that E should be $< V(r_f)$. Choice of r_f in this case is more difficult. One should ensure that $|u(r_f)/u_{\max}| \ll 1$, where $u_{\max}$ is the maximum of the wave function, obtained after an initial estimate.
- If the wave function at every r-mesh point is stored in the current energy-cycle, step (5) can be dispensed with. We can take the wave functions of the two subintervals, make them continuous (by multiplying one of them by an appropriate constant) and normalize it over the entire chosen interval $[r_0, r_f]$.
- Use of Newton–Raphson method to solve Eq. (10.2) will be fast and without problem, if the energy eigenvalues are far apart from each other. However, if they are closely spaced, problem may arise from different branches of the $G(E)$ versus E curve (see below). In this case, we can force a slower energy-convergence,

replacing ΔE by $\nu \Delta E$, with ν chosen to be a small fraction, such that $E + \nu \Delta E$ belongs to the same branch of the curve. This will assure convergence to the desired eigenvalue, although it will be slower.

- One can get the radial excitation quantum number by counting the nodes in $u(r)$. In general, there will be no difficulty in calculating $D_L(E)$ and $D_R(E)$, even if $u(r)$ has one or more nodes. However, these will diverge if r_m happens to be a node of $u(r)$. In this case, one can simply change r_m. Note that the solution is independent of the choice of r_m. But it should be so chosen that $|u(r_m)|$ is large, in order that the numerical errors in calculation of log-derivatives, and hence in E, are small. Also, we should avoid an extremum of $u(r)$ at r_f, where the log-derivative vanishes. Hence, a suitable choice for r_m will be a point which is somewhat on the right of the minimum of $V(r) + \frac{\hbar^2}{2\mu} \frac{l(l+1)}{r^2}$, close to which $|u(r)|$ has its maximum.
- For a potential which can support several bound states, a plot of $G(E)$ against E shows several continuous branches, separated by infinite discontinuities at specific E values $E_a, E_b, \ldots$ and zeros at $E_1, E_2, \ldots$ The latter are the eigenvalues, the smallest being the ground state (no nodes), the next one the first excited state (one node), and so on. At $E = E_a$, $E = E_b$, etc., $u(r_m)$ has a node.
- In view of the above, an efficient way to get all energy eigenvalues is to calculate $G(E)$ at $E = V_0 + (i - 1)h_E$ with $i = 1, 2, \ldots$, where V_0 is the minimum of $V(r) + \frac{\hbar^2}{2\mu} \frac{l(l+1)}{r^2}$ and h_E is the energy mesh interval. $G(E)$ is scanned for change of sign, as E is increased. When a sign change occurs, current value of h_E is divided by 10 and scanning of $G(E)$ is done again, starting from the last E-mesh value (corresponding to h_E before it was divided by 10). This process is stopped when $h_E < \epsilon$. The last E value is the corresponding energy eigenvalue. For the next excited state, a new scanning is then started from $E + \epsilon$ with the original h_E, untill all desired eigenvalues are obtained. Spurious solutions corresponding to infinite discontinuities of $G(E)$ are eliminated by skipping the energy value, when $|G(E)|$ *increases rapidly* with successive reduction of h_E. For a true solution E_n corresponding to $G(E_n) = 0$, the quantity $|G(E)|$ will gradually *decrease* with decreasing h_E.

10.2   Solution of Coupled Differential Equations

The Schrödinger equation reduces to a set of CDE for a system containing more than two particles, as we saw in Chaps. 3 and 4. The CDE also results for a single particle or for the two-body system, when the potential is noncentral. Thus for an exact numerical solution of the Schrödinger equation in such cases, a numerical algorithm for the solution of CDE is necessary. In this section, we discuss such an algorithm, called the renormalized Numerov (RN) method, following Ref. [4].

We consider, the general form of CDE, given by Eqs. (3.30) and (4.34), rewriting it in a general form

$$\left(-\frac{\hbar^2}{m}\frac{d^2}{d\xi^2} - E\right)u_\kappa(\xi) + \sum_{\kappa'} V_{\kappa,\kappa'}(\xi)u_{\kappa'}(\xi) = 0, \quad (\kappa = 1, 2, \ldots, M). \quad (10.3)$$

Here, a single cardinal number index κ ($= 1, 2, \ldots$) represents the combination of all relevant quantum numbers in a chosen fixed sequence. For the chosen truncation of the set of CDE, the maximum value of κ is M. The $M \times M$ potential matrix V includes the hypercentrifugal term and any hypercental term present (as diagonal elements).

10.2.1 Exact Solution of the CDE

We rewrite Eq. (10.3) in a matrix form

$$\left([I]\frac{d^2}{d\xi^2} + [Q(\xi)]\right)[u(\xi)] = [0], \quad (10.4)$$

where a symbol enclosed in square brackets represents an $M \times M$ matrix. The matrices $[I]$ and $[0]$ represent the $M \times M$ unit and null matrices respectively. These notations will be followed throughout this section. The matrix $[Q(\xi)]$ is defined as

$$[Q(\xi)] = \frac{m}{\hbar^2}\left(E[I] - [V(\xi)]\right). \quad (10.5)$$

Note that the matrices $[V]$, $[Q]$, and $[u]$ are functions of ξ. It is also to be noted that before application of boundary conditions, Eq. (10.3) has M linearly independent solutions. These are arranged along the columns of the $M \times M$ matrix $[u]$. Application of the boundary conditions select a particular linear combination of the columns as the eigen vector (see below).

By the same criteria as in Sect. 10.1, we replace end points of the semi-infinite interval $0 \leq \xi < \infty$ by the finite interval $\xi_0 \leq \xi \leq \xi_f$, with sufficiently small, but nonzero, ξ_0 and sufficiently large, but finite, ξ_f. Next introduce $(n + 1)$ equi-spaced grid points $\xi_0, \xi_1, \xi_2, \ldots, \xi_n (= \xi_f)$ in this interval, with mesh size $p = (\xi_f - \xi_0)/n$, such that $\xi_j = \xi_0 + jp$, $(j = 0, n)$. We also introduce short-hand notations $[Q_j] \equiv [Q(\xi_j)]$, $[u_j] \equiv [u(\xi_j)]$, etc., which are $M \times M$ matrices at the jth grid point ξ_j.

There are several algorithms for the point-by-point integration of a single second order differential equation. One of them is a simple three-term recurrence formula called the Numerov algorithm [5], which follows directly from the definition of second derivative of a function. This can be adopted for matrix equation (10.4) as

$$\Big([I] - [T_{j+1}]\Big)[u_{j+1}] - \Big(2[I] + 10[T_j]\Big)[u_j] + \Big([I] - [T_{j-1}]\Big)[u_{j-1}] = [0], \quad (10.6)$$

where $[T_j] = -\frac{p^2}{12}[Q_j] = -\frac{p^2}{12}[Q(\xi_j)]$.

For the RN algorithm, we define

$$[F_j] = \Big([I] - [T_j]\Big)[u_j]. \tag{10.7}$$

Substituting in Eq. (10.6) one gets

$$[F_{j+1}] - [W_j][F_j] + [F_{j-1}] = [0], \tag{10.8}$$

where

$$[W_j] = \Big([I] - [T_j]\Big)^{-1}\Big(2[I] + 10[T_j]\Big). \tag{10.9}$$

Next a two-term recurrence relation is obtained by introducing the matrix

$$[R_j] = [F_{j+1}][F_j]^{-1}. \tag{10.10}$$

Post-multiply Eq. (10.8) by $[F_j]^{-1}$ and use Eq. (10.10) to get

$$[R_j] = [W_j] - [R_{j-1}]^{-1}. \tag{10.11}$$

This is the basic two-term recurrence relation of the RN method. Knowing $[R_i]$ and using Eq. (10.11) one can calculate $[R_{i+1}]$, $[R_{i+2}]$, ... in the forward direction. For the simple choice of initial condition $[u_0] = [0]$ and $[u_1] \neq [0]$ (which are equivalent to $[u_0] = [0]$ and $[u'_0] \neq [0]$), we get from Eq. (10.7) $[F_0] = [0]$ and $[F_1] \neq [0]$. Then from Eq. (10.10), $[R_0]^{-1} = [F_0][F_1]^{-1} = [0]$. With this starting value, the forward recurrence relation, Eq. (10.11) can be used repeatedly to calculate $[R_1]$, $[R_2]$, ...

A similar recurrence relation for backward integration can be found by defining a matrix

$$[\hat{R}_j] = [F_{j-1}][F_j]^{-1}. \tag{10.12}$$

Substitution of this definition in Eq. (10.8) gives

$$[\hat{R}_j] = [W_j] - [\hat{R}_{j+1}]^{-1}. \tag{10.13}$$

Knowing $[\hat{R}_i]$, one can calculate $[\hat{R}_{i-1}]$, $[\hat{R}_{i-2}]$, ... using the above backward recurrence relation. One can start the backward integration grid-by-grid from the end point ξ_n, assuming that $[u(\xi_n)] = [0]$ and $[u(\xi_{n-1})] \neq [0]$, which gives $[\hat{R}_n]^{-1} = [0]$.

We next choose a matching grid point q, such that ξ_q is larger than the position of the minimum of $V_{11}(\xi)$ (assumed to be the most dominant). The continuity condition is the matching of log-derivatives of each partial wave, as obtained by integrating the CDE forward and backward to the match point q. Alternative requirement is to match the partial waves at two consecutive grid points q (integrating forward) and $q + 1$ (integrating backward). For this, we integrate the CDE grid-by-grid in forward direction using Eq. (10.11), starting from 0 to the matching grid point q, and in backward direction using Eq. (10.13) from n to $M + 1$. Thus, we get $[R_q]$ and $[\hat{R}_{q+1}]$ and the matching condition is

$$\left([R_q] - [\hat{R}_{q+1}]^{-1}\right) f(\xi_q) = 0, \tag{10.14}$$

where

$$f(\xi_q) = \left([I] - [T_q]\right) u(\xi_q) \tag{10.15}$$

is the column vector of eigen functions at the match point, being the appropriate linear combination of the columns of $[u_q]$.

Solution of Eq. (10.14) gives the energy eigenvalue E, since $[R_q]$ and $[\hat{R}_{q+1}]$ depend on E. Since Eq. (10.14) is a homogeneous equation, it will have a nontrivial solution only if the determinant of the matrix vanishes

$$D(E) = \left|\left([R_q] - [\hat{R}_{q+1}]^{-1}\right)\right| = 0. \tag{10.16}$$

With the correct energy E satisfying Eq. (10.16), Eq. (10.14) is solved for the column vector $f(\xi_q)$ subjected to normalization. The complete eigen function in the two subintervals is then obtained by iteration in the opposite direction from the match point

$$f(\xi_i) = [R_i]^{-1} f(\xi_{i+1}), \qquad i = q - 1, q - 2, \ldots, 0, \tag{10.17}$$

for the first sub-interval, and

$$f(\xi_i) = [\hat{R}_i]^{-1} f(\xi_{i-1}), \qquad i = q + 1, q + 2, \ldots, n, \tag{10.18}$$

for the second sub-interval.

The standard boundary condition at the end grid point n is taken as $[\hat{R}_n]^{-1} = [0]$. This assumes that the wave function of the end grid point vanishes, while that at the previous grid point is nonvanishing [see Eqs. (10.12) and (10.7)]. It can be improved using exact solution of the CDE, Eq. (10.4), in the asymptotic region where $[V] = [0]$, so that the CDE becomes decoupled and each component is proportional to $\exp(-\beta\xi)$ with $\beta = \sqrt{-\frac{mE}{\hbar^2}}$ [6]. This gives

$$[\hat{R}_n]^{-1} = e^{-\beta p}[I]. \tag{10.19}$$

Use of Eq. (10.19) as a boundary condition over the standard one gives a 10 % improvement in the deuteron binding energy with Reid soft core (RSC) potential [6]. A similar modification for the initial boundary value $[R_0]^{-1}$ at ξ_0 can be obtained by using the $\xi \to 0$ analytical solution of the CDE.

Algorithm for the RN Method

From the procedure discussed above, the algorithm for the RN method is given by the following steps:

1. Choose a trial energy (E), initial increment (ΔE), precision required in energy calculation (ϵ), end points of the chosen interval in ξ variable (ξ_0, ξ_f), number of mesh intervals (n) and match grid point (q).
   Calculate $p = (\xi_f - \xi_0)/n$. Set IND = 0, $D = 0$.
2. Set IND = IND + 1 and $D_0 = D$
   Calculate initial $[R_0]^{-1}$ and final $[\hat{R}_n]^{-1}$ for starting the iterations.
3. Use Eq. (10.11) to iterate $[R_i]$, point-by-point from $i = 1$ to $i = q$. Save $[R_q]$
4. Use Eq. (10.13) to iterate $[\hat{R}_i]$, point-by-point from $i = n$ to $i = q + 1$. Save $[\hat{R}_{q+1}]$.
5. Calculate the determinant $D = \left| \left([R_q] - [\hat{R}_{q+1}]^{-1} \right) \right|$
6. For IND = 1:

   (a) Replace E by $E + \Delta E$.
   (b) Go to step (2).

7. For IND > 1:

   (a) calculate $\frac{dD}{dE} = \frac{D - D_0}{\Delta E}$, Then calculate $\Delta E = -\frac{D}{\frac{dD}{dE}}$.
   (b) Replace E by $E + \Delta E$.
   (c) If $|\frac{\Delta E}{E}| < \epsilon$ go to step (8).
       Otherwise, go to step (2)

8. Final eigen energy = E.
9. Set the first component of $f(\xi_q)$ vector equal to 1 and solve Eq. (10.14), which now is a set of $(M - 1)$ linear inhomogeneous equations (LIE).
10. Starting from this column vector $f(\xi_q)$, use Eq. (10.17) with $[R_i]$ recalculated for the final energy, to iterate backward and calculate column vectors $f(\xi_i)$ with $i = q - 1, q - 2, \ldots, 0$, storing each component of the column vector $f_\nu(\xi_i)$, $\nu = 1, \ldots, M$.
11. Starting from the column vector $f(\xi_q)$, use Eq. (10.18) with $[\hat{R}_i]$ recalculated for the final energy, to iterate forward and calculate column vectors $f(\xi_i)$ with $i = q+1, q+2, \ldots, n$, storing each component of the column vector $f_\nu(\xi_i)$, $\nu = 1, \ldots, M$
12. Normalize the wave function according to

$$\int_{\xi_0}^{\xi_n} \sum_{\nu=1}^{M} |f_\nu(\xi)|^2 d\xi = 1.$$

In the above, we used the simple Newton–Raphson method [2] for solving the energy equation, as an illustration. It can be replaced by more sophisticated root finding algorithms discussed in Sect. 10.1. Suitable standard algorithms for solving the LIE, calculation of determinant and numerical integration should be chosen [2, 3]. Convergence behavior of the numerical solution of a system of CDE by the RN method has been studied in Ref. [7].

10.2.2 Introduction of Hypercentral Average

Let us consider, the general form of the CDE

$$
\left(-\frac{\hbar^2}{m}\frac{d^2}{d\xi^2} + \frac{\hbar^2}{m}\frac{\mathcal{L}_\kappa(\mathcal{L}_\kappa + 1)}{\xi^2} + U(\xi) - E \right) u_\kappa(\xi)
$$
$$
+ \sum_{\kappa'} V_{\kappa,\kappa'}(\xi) u_{\kappa'}(\xi) = 0, \quad (\kappa = 1, 2, \ldots, M), \tag{10.20}
$$

in which the hypercentrifugal term, as also a possible hypercentral potential $U(\xi)$, have been kept explicitly outside the coupling potential V. It is obvious that convergence of energy and expansion of the wave function will be faster if magnitude of the off-diagonal matrix elements of V are small compared to magnitude of the hypercentral terms. Hence for an arbitrary interaction potential $V_{ij} \equiv V(\vec{r}_i - \vec{r}_j)$, the convergence will be faster, if we calculate a hypercentral average $V_0(\xi)$ (which is the average of V_{ij} over all hyperangles) and replace V_{ij} by $V_{ij} - V_0(\xi)$, while adding $V_0(\xi)$ to the diagonal term. Thus, rate of convergence and therefore accuracy of the numerical calculation can be improved by separating the hypercentral average [8]. The hypercentral average for a general two-body potential $V(\vec{r}_i - \vec{r}_j)$ is given by

$$
V_0(\xi) = \int V(\vec{r}_i - \vec{r}_j) d\Omega_{3N}. \tag{10.21}
$$

Hence Eq. (10.20) is replaced by

$$
\left(-\frac{\hbar^2}{m}\frac{d^2}{d\xi^2} + \frac{\hbar^2}{m}\frac{\mathcal{L}_\kappa(\mathcal{L}_\kappa + 1)}{\xi^2} + U(\xi) + \frac{A(A-1)}{2}V_0(\xi) - E \right) u_\kappa(\xi)
$$
$$
+ \sum_{\kappa'} \overline{V}_{\kappa,\kappa'}(\xi) u_{\kappa'}(\xi) = 0, \quad (\kappa = 1, 2, \ldots, M), \tag{10.22}
$$

where $\overline{V}$ is given by

$$
\overline{V} = \sum_{i<j=2}^{A} \left(V(\vec{r}_i - \vec{r}_j) - V_0(\xi) \right). \tag{10.23}
$$

It is the *residual potential* after subtraction of the hypercentral average. Note that in Eq. (10.22) the compensation term has a factor $A(A - 1)/2$, corresponding to the number of interacting pairs.

Inclusion of the hypercentral average improves the rate of convergence considerably. It would be a fairly good approximation for most two-body potentials, even if we disregard the coupling term altogether in Eq. (10.22).

Since the integro-differential equation (Chap. 9) was derived from the CDE for the PHEM, a similar hypercentral average $V_0(\xi)$ can be subtracted from the two-body potential and $\frac{A(A-1)}{2} V_0(\xi)$ added to the left side of Eq. (9.30).

10.2.3 Hyperspherical Adiabatic Approximation

To reduce the bulk of computation in solving the CDE, one can use the hyperspherical adiabatic approximation (HAA) for decoupling the CDE adiabatically into a single differential equation [9]. Justification of this approximation is based on a physical criterion, viz., the assumption that hyperradial motion is slow compared to hyperangular motion. This assumption is likely to be valid, since hyperradial excitations correspond to breathing modes. Hence, one can decouple hyperradial motion adiabatically and solve hyperangular motion for a fixed hyperradius (ξ) to get an effective potential as a parametric function of ξ. The hyperradial equation with this potential is then solved to get energy of the system. This process is similar in spirit to the Born–Oppenheimer approximation (BOA), in which motion of nuclei (heavy particles) in an atom is assumed to be slow compared with that of electrons (light particles) and these two motions are separated adiabatically [11]. First, the Schrödinger equation of the electrons is solved for a fixed configuration of the nuclei. This gives the potential energy of that particular nuclear configuration, which is used as the effective potential in which the nuclei move. A simple illustrative example is the hydrogen molecular ion H_2^+, in which motion of the electron is solved for the protons kept at a fixed separation, giving the effective inter-protonic potential. Relative motion of the two protons is then solved with this potential, which gives the energy of the system [10, 11]. Justification of the assumption in BOA has a stronger physical basis than that in HAA. Still in actual atomic and nuclear calculations, HAA appears to be surprisingly accurate [9, 13]. It has been shown to have a mathematical basis [12] for a smooth interaction potential.

The general CDE in ξ variable, Eq. (4.34), is obtained by substituting the HH expansion of A-body relative wave function (4.17) appearing in the N-body Schrödinger equation, followed by projection on a particular HH. We can express the same equation in terms of hyperspherical variables ξ and Ω_D by writing Eq. (4.17) as $\Psi = \xi^{-(D-1)/2}\Phi(\xi, \Omega_D)$ (where $D = 3N$) and substituting in the N-body Schrödinger equation (4.5), with kinetic energy operator given by Eq. (4.8)

$$\left[-\frac{\partial^2}{\partial \xi^2} + \frac{\hat{\mathcal{L}}_D^2(\Omega_D) + (D-1)(D-3)/4}{\xi^2} + v(\xi, \Omega_D) + k^2 \right]\Phi(\xi, \Omega_D) = 0,$$

(10.24)

where $v(\xi, \Omega_D) = \frac{m}{\hbar^2}V(\xi, \Omega_D)$ and $k^2 = -\frac{mE}{\hbar^2}$. In the HAA one assumes hyperradial motion to be slow compared to hyperangular motion, so that for a fixed value of ξ, one can solve the Ω_D motion adiabatically, i.e.,

$$\left[\frac{\hat{\mathcal{L}}_D^2(\Omega_D) + (D-1)(D-3)/4}{\xi^2} + v(\xi, \Omega_D) \right]B_\lambda(\xi, \Omega_D) = \omega_\lambda(\xi)B_\lambda(\xi, \Omega_D).$$

(10.25)

Naturally, the eigenvalue ω_λ and the eigen function $B_\lambda(\Omega_D)$ are parametric functions of ξ. The set of eigen functions $\{B_\lambda(\xi, \Omega_D)\}$, being the solution of the Hermitian differential operator within the square brackets in Eq. (10.25), forms a complete set for expansion of an arbitrary function of Ω_D. This set is called the *adiabatic subset*. Hence $\Phi(\xi, \Omega_D)$ (for a fixed ξ) can be expanded as

$$\Phi(\xi, \Omega_D) = \sum_\lambda \zeta_\lambda(\xi)B_\lambda(\xi, \Omega_D).$$

(10.26)

Next, $B_\lambda(\xi, \Omega_D)$ can be expanded in the set of HH

$$B_\lambda(\xi, \Omega_D) = \sum_{L',[L']} \chi_{(L',[L']),\lambda}(\xi)\mathcal{Y}_{L',[L']}(\Omega_D).$$

(10.27)

Substituting Eq. (10.27) in (10.25), premultiplying by $\mathcal{Y}_{L,[L]}(\Omega_D)^*$ and integrating over $d\Omega_D$, we have

$$\sum_{(L',[L'])} \left[\frac{\mathcal{L}_L(\mathcal{L}_L + 1)}{\xi^2}\delta_{(L,[L]),(L',[L'])} + \langle L, [L]|v|L', [L']\rangle \right]\chi_{(L',[L']),\lambda}(\xi)$$

$$= \omega_\lambda(\xi)\chi_{(L,[L]),\lambda}(\xi).$$

(10.28)

Here the HH $\mathcal{Y}_{L,[L]}(\Omega_D)$ is also the eigenfunction of the operator $\hat{\mathcal{L}}_D^2(\Omega_D) + (D-1)(D-3)/4$ corresponding to eigenvalue $\mathcal{L}_L(\mathcal{L}_L + 1)$, with $\mathcal{L}_L = L + \frac{D-3}{2}$. For a fixed ξ, Eq. (10.28) is a matrix eigenvalue equation. For a given λ, the real numbers $\chi_{(L,[L]),\lambda}(\xi)$ are the elements of the eigen column vector, satisfying the orthonormalization

$$\sum_{L,[L]} \chi_{(L,[L]),\lambda}(\xi)\chi_{(L,[L]),\lambda'}(\xi) = \delta_{\lambda,\lambda'}. \qquad (10.29)$$

If we expanded the relative wave function of the $A = N + 1$ body system directly in the HH basis

$$\Phi(\xi, \Omega_D) = \sum_{L,[L]} u_{L,[L]}(\xi)\mathcal{Y}_{L,[L]}(\Omega_D), \qquad (10.30)$$

then its substitution in the many-body relative Schrödinger equation gives

$$\left[-\frac{d^2}{d\xi^2} + \frac{\mathcal{L}_L(\mathcal{L}_L + 1)}{\xi^2} + k^2 \right] u_{L,[L]}(\xi) + \sum_{L',[L']} \langle L, [L]|v|L', [L']\rangle u_{L',[L']}(\xi) = 0.$$

$$(10.31)$$

Comparison of Eq. (10.30) with Eqs. (10.26) and (10.27) gives

$$u_{L,[L]}(\xi) = \sum_{\lambda} \zeta_{\lambda}(\xi)\chi_{(L,[L]),\lambda}(\xi). \qquad (10.32)$$

Substituting Eq. (10.32) in (10.31), using Eq. (10.28) and taking inner product with $\chi_{(L,[L]),\lambda}(\xi)$, we get

$$\left[-\frac{d^2}{d\xi^2} + \omega_{\lambda}(\xi) + k^2 \right] \zeta_{\lambda}(\xi) - 2 \sum_{\lambda',(L,[L])} \frac{d\zeta_{\lambda'}(\xi)}{d\xi} \left(\chi_{(L,[L]),\lambda}(\xi) \frac{d\chi_{(L,[L]),\lambda'}(\xi)}{d\xi} \right)$$

$$- \sum_{\lambda',(L,[L])} \zeta_{\lambda'}(\xi) \left(\chi_{(L,[L]),\lambda}(\xi) \frac{d^2\chi_{(L,[L]),\lambda'}(\xi)}{d\xi^2} \right) = 0. \qquad (10.33)$$

Differentiating Eq. (10.29) once with respect to ξ, we have

$$\sum_{L,[L]} \left(\chi_{(L,[L]),\lambda}(\xi) \frac{d\chi_{(L,[L]),\lambda'}(\xi)}{d\xi} + \chi_{(L,[L]),\lambda'}(\xi) \frac{d\chi_{(L,[L]),\lambda}(\xi)}{d\xi} \right) = 0. \qquad (10.34)$$

Differentiating once again for $\lambda = \lambda'$, we have

$$\sum_{L,[L]} \chi_{(L,[L]),\lambda}(\xi) \frac{d^2\chi_{(L,[L]),\lambda}(\xi)}{d\xi^2} = -\sum_{L,[L]} \left| \frac{d\chi_{(L,[L]),\lambda}(\xi)}{d\xi} \right|^2 \qquad (10.35)$$

Substituting Eq. (10.34) for $\lambda = \lambda'$ and Eq. (10.35) in (10.33), we have

$$\left[-\frac{d^2}{d\xi^2} + \omega_\lambda(\xi) + k^2 + \sum_{L,[L]} \left| \frac{d\chi_{(L,[L]),\lambda}(\xi)}{d\xi} \right|^2 \right] \zeta_\lambda(\xi)$$

$$- \sum_{\lambda' \neq \lambda} \left[2\frac{d\zeta_{\lambda'}(\xi)}{d\xi} \left(\sum_{L,[L]} \chi_{(L,[L]),\lambda}(\xi) \frac{d\chi_{(L,[L]),\lambda'}(\xi)}{d\xi} \right) \right.$$

$$\left. + \zeta_{\lambda'}(\xi) \left(\sum_{L,[L]} \chi_{(L,[L]),\lambda}(\xi) \frac{d^2\chi_{(L,[L]),\lambda'}(\xi)}{d\xi^2} \right) \right] = 0. \qquad (10.36)$$

Up to this point, there is no approximation and Eq. (10.36) is equivalent to the exact CDE, Eq. (10.31). For weak hyperangular dependence of the potential $v(\xi, \Omega_D)$ [this can be enhanced by subtracting and adding the hypercentral average, see Sect. 10.2.2], the off-diagonal elements $\langle L, [L]|v|L', [L']\rangle$ (with $L' \neq L$) are small and Eq. (10.28) shows that the column vectors $\chi_{(L,[L]),\lambda}(\xi)$ are very slowly varying functions of ξ. Hence its first and second derivatives are even smaller. Thus the coupling terms in Eq. (10.36) will be quite small. If we disregard the coupling terms (sum over $\lambda' \neq \lambda$), the resulting equation is called the uncoupled adiabatic approximation (UAA). The fourth term is always positive and represents the adiabatic overbinding correction term. Disregarding even this term, we get the extreme adiabatic approximation (EAA), which gives overbinding [14]. For the ground state one uses the *lowest eigen potential* $\omega_0(\xi)$ for $\lambda = 0$ and the UAA becomes

$$\left[-\frac{d^2}{d\xi^2} + \omega_0(\xi) + k^2 + \sum_{L,[L]} \left| \frac{d\chi_{(L,[L]),0}(\xi)}{d\xi} \right|^2 \right] \zeta_0(\xi) = 0. \qquad (10.37)$$

While the EAA overbinds, the UAA underbinds, satisfying an energy inequality relation [14]

$$E_{\text{EAA}} \leq E_{\text{exact}} \leq E_{\text{UAA}}, \qquad (10.38)$$

the exact energy E_{exact} being closer to E_{UAA}.

Applicability of HAA

The HAA is clearly advantageous, as it reduces the difficult numerical task of solving a large system of CDE into that of solving a single differential equation. The question is: when is the HAA applicable? The physical criterion is the assumption that the hyperradial motion is slow compared to the hyper-angular motion. However in most physical situations, this cannot be guaranteed a priori. We can check the accuracy of HAA only in test cases, where exact results are known (see below). But we can enhance its applicability by subtracting and adding the hypercentral average from the many-body potential $v(\xi, \Omega_D)$, as discussed above. Even then, its accuracy in diverse (nuclear, atomic, and molecular systems) test cases is better than 1 % (see below), which is quite intriguing. It has been shown that there is a mathematical justification for the surprising accuracy of the HAA for smooth potentials [12]. The HAA becomes exact in the $\xi \to 0$ and the asymptotic limits, whereas for intermediate values of ξ, it is good if the potential does not change too rapidly in any region. The disregarded coupling terms become small, if eigenpotentials are fairly widely separated. When this is satisfied, the accuracy of the HAA is quite high. This is in contrast with the BOA, whose accuracy depends on the validity of the assumption that the heavy particles move slowly compared to the light particles.

Accuracy of HAA

As we have indicated above, the HAA enjoys a surprisingly high accuracy for widely different systems, *viz*, nuclear, atomic, or molecular systems. It has been tested for the trinucleon (^{3}H and ^{3}He nuclei), atomic (several two-electron atoms) and molecular (H_2^+ ion) three-body systems against exact solutions of up to 24 coupled equations. In nuclear systems, the nucleon–nucleon (N-N) potential was chosen to be semi-realistic soft-core potentials, with the core part having up to a fairly strong repulsive core, as in the S3 potential. Thus the N-N potential changes rapidly at short separations for the S3 potential, but still the accuracy of UAA is better than 1 % [9]. For the atomic and molecular systems, the basic interaction is Coulomb. Deviations of UAA from exact ground state energies are less than 0.4 % [13] for two-electron systems like H^-, He, Li^+, Be^{2+}, B^{3+}, $ee\mu$, and Ps^-, with up to 24 CDEs. Even for the first excited state of neutral Helium atom, the deviation is only about 0.67 %. For the simple hydrogen molecular ion H_2^+, the deviation is about 0.4 % [12]. Thus the HAA appears to be a very good approximation scheme for most few-body systems.

Numerical Procedure for UAA

The final UAA equation for the ground state, Eq. (10.37), is a single differential equation, which can be solved by the method of Sect. 10.1. We need to calculate the eigenpotential and the overbinding correction term. The procedure can be summarized as follows.

1. Select a maximum L value L_{max}. Set up a single index κ for the combination $(L, [L])$. κ_{max} corresponds to L_{max}.
2. For the chosen interval $[\xi_I, \xi_F]$ with N uniform mesh-intervals, set up mesh points ξ_i, $i = 1, N + 5$, where $\xi_i = \xi_I + (i - 5)h$ with $h = (\xi_F - \xi_I)/N$. Start a loop over $i = 1, N + 5$. We include four additional mesh points before the chosen initial point ξ_I, in order to calculate the first derivative of the eigenvector from ξ_I (see step 5 below).
3. For a particular ξ_i, calculate potential matrix $\langle L, [L] | v(\xi_i, \Omega_D) | L', [L'] \rangle$ (see Sect. 10.3).
4. Set up the $\kappa_{max} \times \kappa_{max}$ matrix

$$M_{\kappa, \kappa'} = \left[\frac{\mathcal{L}_L(\mathcal{L}_L + 1)}{\xi_i^2} \delta_{(L, [L]), (L', [L'])} + \langle L, [L] | v(\xi_i, \Omega_D) | L', [L'] \rangle \right],$$

and diagonalize it to obtain lowest eigenvalue $\omega_0(\xi_i)$ and corresponding column eigenvector $\{\chi_{\kappa, 0}(\xi_i), \ \kappa = 1, \kappa_{max}\}$. Store these quantities in suitable arrays.

5. For $i \geq 5$, compute the first derivative of $\chi_{\kappa,0}(\xi_i)$, $(\kappa = 1, \kappa_{max})$ employing a five-point formula [15], using already calculated and stored $\chi_{\kappa,0}$ values. Next compute the overbinding correction term

$$\sum_{\kappa=1}^{\kappa_{max}} \left| \frac{d\chi_{\kappa,0}(\xi)}{d\xi} \big|_{\xi=\xi_i} \right|^2$$

6. Include $\omega_0(\xi_i)$ and the overbinding term of step 5 in Eq. (10.37) and solve for E_{UAA} and $\zeta_0(\xi)$ by the method of Sect. 10.1 in the interval $[\xi_I, \xi_F]$. The eigenfunction is normalized according to

$$\int_{\xi_I}^{\xi_F} |\zeta_0(\xi)|^2 d\xi = 1.$$

Partial waves of original Eq. (10.31) are given approximately as

$$u_\kappa(\xi_i) \approx \zeta_0(\xi_i)\chi_{\kappa,0}(\xi_i).$$

10.3 Computation of Matrix Element

Computation of all the matrix elements $\langle L, [L] | v(\xi_i, \Omega_D) | L', [L'] \rangle$ can be reduced to a minimum set by using the symmetries of the chosen basis states (see Chaps. 4 and 5). Next, most of the angular integrations can be done analytically, leaving a fewer dimensional (one dimensional for the three-body system) integral to be done numerically. In most cases (where the potential has a smooth dependence on its arguments) such integrals can be computed by standard quadratures [3]. Alternately, if the size of the chosen basis set is large, one can use the geometrical structure coefficients (see Chaps. 3, 5 and 7) to reduce the bulk of computation. The geometrical structure coefficients can be computed in an elegant fashion using the linear inhomogeneous equation method (see Chaps. 3 and 5).

Special discussion is necessary for computation of matrix elements if the integrand depends strongly on its arguments over some domain of integration. An example is provided by the matrix elements in the correlated potential harmonic basis, Eq. (8.27), for a large number of particles. In that case, the weight function $W_l(z) = (1 - z)^\alpha (1 + z)^\beta$ is appreciable only in a tiny interval near the lower limit of integration, over which it varies extremely rapidly (see Chap. 8). Unless special care is taken, a standard numerical quadrature will miss the major contribution to the integral. An accurate computation of the integral can be done by subdividing the original interval $[a, b]$ of integration into subintervals, which are sufficiently dense over the domain giving the major contribution and using standard quadratures for each subinterval. For a Bose condensate containing a large number of particles, the major contribution to the integral in Eq. (8.27) comes from near the lower limit of integration $z = -1$. Then the interval $[-1, 1]$ is subdivided into n gradually increasing subintervals: $h_0, ch_0, c^2 h_0, \ldots, c^{n-1} h_0$ (with c a constant > 1), such that [16]

$$h_0 + ch_0 + c^2 h_0 + \cdots + c^{n-1} h_0 = b - a = 2$$

which gives

$$h_0 = \frac{2(c-1)}{(c^n - 1)}.$$

The constant $c > 1$ is so chosen that the first subinterval h_0 is desirably small for the particular case. The integral of Eq. (8.27) is then replaced by a sum of n subintegrals:

$$\int_{-1}^{1} = \int_{-1}^{-1+h_0} + \int_{-1+h_0}^{-1+h_0+ch_0} + \cdots + \int_{1-c^{n-1}h_0}^{1} .$$

Each subintegral is evaluated by a standard quadrature, e.g., the 32-point Gauss–Legendre quadrature. Increasing c by a small amount, the first subinterval can be made extremely small even for a relatively small number of subintervals. For example, fixing $n = 20$, the first subinterval h_0 can be reduced from about 2×10^{-6} to $\sim 10^{-14}$ by increasing c from 2 to 5. Thus even for very large particle number (large α), the integral can be evaluated fast and accurately, using a small number of subintervals.

References

1. Reed, B.C.: Quantum Mechanics. Jones and Bartlett Publishers, London (2010). First India Edition
2. Rajaraman, V.: Computer oriented Numerical Methods, 3rd edn. Prentice-Hall of India, New Delhi (1993)
3. Press, W.H., Teukolsky, S.A., Vetterling, W.T., Flannery, B.P.: Numerical Recipes in FORTRAN The Art of Scientific Computing, 2nd edn. Cambridge University Press, Cambridge (1992)
4. Johnson, B.R.: J. Chem. Phys. **69**, 4678 (1978)
5. Blatt, J.M.: J. Comput. Phys. **1**, 382 (1967)
6. Ghosh, A.K., Das, T.K.: Fizika **22**, 521 (1990)
7. Ghosh, A.K., Das, T.K.: Fizika **19**, 103 (1987)
8. Adam, R.M., Sofianos, S.A.: Phys. Rev. A **82**, 053635 (2010)
9. Das, T.K., Coelho, H.T., Fabre de la Ripelle, M.: Phys. Rev. C **26**, 2281 (1982)
10. Schiff, L.I.: Quantum Mechanics, 3rd edn. Mc Graw Hill Co, Singapore (1968)
11. Konishi, K., Paffuti, G.: Quantum Mechanics: A New Introduction, 1st edn. Oxford University Press, UK (2009)
12. Brito, V.P., Coelho, H.T., Das, T.K.: Phys. Rev. A **40**, 3346 (1989)
13. Adhikari, S.K., Brito, V.P., Coelho, H.T., Das, T.K.: Nuo. Cim. B **107**, 77 (1992); Das, T.K., Coelho, H.T., Brito, V.P.: Phys. Rev. C **48**, 2201 (1993); Chattopadhyay, R., Das, T.K.: Phys. Rev. A **56**, 1281 (1997); Das, T.K., Chakrabarti, B.: Int. J. Mod. Phy. A **19**, 4973 (2004)
14. Coelho, H.T., Hornos, J.E.: Phys. Rev. A **43**, 6379 (1991)
15. Abramowitz, M., Stegun, I.A. (eds.): Handbook of Mathematical Functions, p. 883. Dover Publications Inc., New York (1970)
16. Das, T.K., Canuto, S., Kundu, A., Chakrabarti, B.: Phys. Rev. A **75**, 042705 (2007)

Index

© Springer India 2016
T.K. Das, *Hyperspherical Harmonics Expansion Techniques*,
Theoretical and Mathematical Physics, DOI 10.1007/978-81-322-2361-0

Printed in the United States
By Bookmasters